AF784151

Just Transitions

Series Editor
Raphael J. Heffron
University of Pau and Pays de l'Adour
Pau, France

Raphael J. Heffron • Victoria R. Nalule
Theophilus Acheampong

Revolution in the Global Critical Minerals Sector

A Law, Economics and Justice Perspective

palgrave
macmillan

Raphael J. Heffron
Abu Dhabi University
Abu Dhabi, United Arab Emirates

Theophilus Acheampong
University of Aberdeen
Aberdeen, UK

Victoria R. Nalule
School of Law
University of Bradford
Bradford, UK

ISSN 2731-6041 ISSN 2731-605X (electronic)
Just Transitions
ISBN 978-3-032-26365-0 ISBN 978-3-032-26366-7 (eBook)
https://doi.org/10.1007/978-3-032-26366-7

© The Author(s), under exclusive license to Springer Nature Switzerland AG 2026

This work is subject to copyright. All rights are solely and exclusively licensed by the Publisher, whether the whole or part of the material is concerned, specifically the rights of translation, reprinting, reuse of illustrations, recitation, broadcasting, reproduction on microfilms or in any other physical way, and transmission or information storage and retrieval, electronic adaptation, computer software, or by similar or dissimilar methodology now known or hereafter developed.
The use of general descriptive names, registered names, trademarks, service marks, etc. in this publication does not imply, even in the absence of a specific statement, that such names are exempt from the relevant protective laws and regulations and therefore free for general use.
The publisher, the authors and the editors are safe to assume that the advice and information in this book are believed to be true and accurate at the date of publication. Neither the publisher nor the authors or the editors give a warranty, expressed or implied, with respect to the material contained herein or for any errors or omissions that may have been made. The publisher remains neutral with regard to jurisdictional claims in published maps and institutional affiliations.

Pattern © Melisa Hasan

This Palgrave Macmillan imprint is published by the registered company Springer Nature Switzerland AG.
The registered company address is: Gewerbestrasse 11, 6330 Cham, Switzerland

If disposing of this product, please recycle the paper.

Foreword

The global energy transition has brought extractives and critical minerals to the centre of economic and geopolitical decision-making. A growing range of strategic and critical minerals underpin electrification, digitalisation and decarbonisation strategies worldwide. In this context, the mining sector has assumed renewed strategic importance within the architecture of the future energy system.

This renewed prominence, however, introduces significant governance and policy challenges.

Critical minerals are geographically concentrated, capital intensive and exposed to geopolitical risk. Supply chains are increasingly shaped by industrial policy, trade realignments and strategic competition. At the same time, environmental, social and governance expectations have intensified. The sector therefore operates at the intersection of climate ambition, industrial strategy, fiscal policy and social accountability.

Ambition without governance breeds instability. Securing critical minerals requires more than political commitment; it requires robust institutions, predictable legal frameworks and transparent fiscal regimes capable of sustaining investment and public trust. Where these foundations are weak, volatility rises, risk premiums increase and development outcomes become uncertain.

The expansion of mineral demand associated with the low-carbon transition heightens these pressures. If extraction is to support sustainable development rather than exacerbate inequality, justice and accountability must be embedded across the full lifecycle of projects—from licensing and taxation to closure and remediation.

It is within this evolving landscape that *Advancing Revolution over Evolution in Law and Policy for Extractives and Critical Minerals Sectors* makes a timely and substantive contribution.

Raphael J. Heffron, Victoria R. Nalule and Theophilus Acheampong argue that incremental adjustments to existing legal frameworks will not suffice. The scale and strategic importance of critical minerals require structural recalibration in law and policy. Their analysis integrates legal doctrine, economic evaluation and governance practice in a manner that reflects the multidimensional realities of the sector.

The JUST Framework developed in this book provides a structured lens through which distributive, procedural, restorative, recognition and cosmopolitan dimensions of justice can be assessed. Importantly, this framework is operational. It speaks directly to fiscal regime design, environmental impact assessment processes, mine closure obligations, benefit-sharing arrangements and social licence to operate. Justice, in this formulation, becomes a practical standard against which extractive governance can be evaluated.

The geopolitical implications of critical minerals are equally significant. Concentration of refining capacity and strategic trade partnerships underscore the limitations of fragmented national approaches. The authors' proposal for a Global Critical Minerals Treaty reflects the need for greater coherence in international governance as supply chains become more interconnected and strategically contested.

Importantly, the authors recognise that mineral wealth is neither inherently a blessing nor a curse. Its developmental impact depends on the strength of institutions, the design of fiscal regimes and the integrity of governance systems. Aligning risk, reward and responsibility remains central to ensuring that mineral endowments translate into sustainable economic outcomes.

The extractive sector is entering a period of structural adjustment driven by climate policy, technological change and geopolitical competition. Legal and policy frameworks must evolve accordingly. The credibility of the energy transition will depend not only on access to critical minerals, but on the legitimacy and resilience of the systems that govern them.

This book contributes meaningfully to that debate. It invites policy-makers, industry leaders and civil society to engage with the legal and normative foundations of extractive governance at a time when economic, environmental and geopolitical stakes are increasingly intertwined.

Crystol Energy
London, UK

Carole Nakhle,

Preface

The importance of the mining sector has been markedly elevated in the twenty-first century as the world accelerates its transition to a low-carbon economy. At the centre of this transformation lies a growing dependence on critical minerals, which are indispensable to renewable energy technologies, battery storage systems and the broader clean energy infrastructure, among others. This book overall explores the background of the industry and its importance. This is interspersed with example internationally and in particular from Africa. The book concludes that, owing to slow progress and the importance of the critical mineral industry, a new UN-supported treaty on critical minerals is needed.

This surge in demand presents significant economic opportunities for resource-rich countries but also highlights deep-rooted global challenges related to governance, equity and sustainability. The global scramble for critical minerals has reignited strategic competition among states, intensified geopolitical tensions and raised pressing questions about justice in the energy transition and the mining sector more broadly.

The justice framework primarily advocates equality and equity in sharing both the benefits and burdens associated with mineral extraction. Therefore, the extraction and governance of mineral resources must transcend economic efficiency and legal justice to encompass social justice and political stability. If managed responsibly, minerals can serve as a lever for peace, security and development. This requires robust legal frameworks promoting transparency and accountability, equitable economic models supporting local development and policies prioritising environmental protection and community rights.

While each country is entitled to exercise sovereign control over its natural resources, international and regional initiatives increasingly play a vital role in promoting justice within the mining sector. This significance arises from the sector's transboundary nature and the security of supply associated with critical minerals. Consequently, international cooperation is indispensable to uphold ethical supply chains and to ensure benefits are distributed fairly, especially to those most affected by mining activities. Initiatives such as the Kimberley Process, the Extractive Industries Transparency Initiative (EITI) and regional frameworks like the African Mining Vision and the Intergovernmental Forum on Mining provide important blueprints for leveraging mineral wealth to foster inclusive development and to minimise resource-driven instability.

This book examines interconnecting legal, economic and normative dimensions of the global mining sector in the age of critical minerals and the energy transition. Employing a multi-disciplinary analysis, the book considers how states, multilateral institutions and private actors might collaborate to ensure that mineral extraction contributes not just to climate goals, but also to broader objectives of justice, peace and sustainable development. By foregrounding these imperatives, the book aims to promote a more equitable and secure global mineral economy that recognises both the promise and the perils of the twenty-first century's resource frontier.

Aberdeen, UK — Raphael J. Heffron
Bradford, UK — Victoria R. Nalule
Abu Dhabi, United Arab Emirates — Theophilus Acheampong

Acknowledgements We would like to extend our heartfelt gratitude to the publisher for their unwavering patience and support throughout the publication process of this book.

Contents

About the Authors

Raphael J. Heffron is a leading global authority on energy justice and an international expert on climate change, sustainability and just transition public policy issues. Currently, he is professor and dean at the College of Law in Abu Dhabi University. Heffron is a qualified Barrister-at-Law, and a graduate of both Oxford (MSc, Christ Church) and Cambridge (MPhil, Darwin, and PhD, Trinity Hall) Universities (UK). He also holds degrees from the University of St. Andrews (MLitt) (UK) and Trinity College Dublin (BA, MA) (Ireland).

His work focuses primarily on achieving a sustainable and just transition to a low-carbon economy, integrating perspectives from governance, management, policy, law and economics. His work has featured in the Top 2% Stanford-Elsevier Scientists List since 2020. He has published over 220 scholarly works across a range of formats and is the world's most cited scholar in the field of energy and just transition law and public policy, with more than 7,000 citations in Scopus and over 13,000 citations in Google Scholar. His publications have been translated into multiple languages including Chinese, Italian, French, Spanish, Arabic and Persian. Heffron has also delivered more than 220 keynote addresses and/or guest lectures in 70 countries worldwide.

Victoria R. Nalule is the Founder and CEO of Nalule Energy and Minerals Consultants (NEM Energy) and Lecturer in Natural Resources Law at the University of Bradford, UK. She holds a PhD in International Energy Law and Policy from the University of Dundee's Centre for Energy, Petroleum and Mineral Law and Policy (CEPMLP). Nalule also serves as an energy arbitrator at the Energy Disputes Arbitration Centre (EDAC) in Turkey. She is the International Chamber of Commerce (ICC) YAAF representative for Africa.

An accomplished author, Nalule has written nine books and numerous scholarly articles focused on energy transitions, climate change, oil, gas, renewable energy and mining. She has provided strategic advice to various governments on high-profile projects, including leading teams in the development of national energy transition strategies, mining policies and extractive industry projects.

Nalule's expertise extends to consulting for leading international organisations, including the African Union, the African Development Bank and the EITI International Secretariat. She actively contributes to global dialogues on energy and mining through media appearances, her YouTube channel and podcast, grounding her reputation as a distinguished and influential international expert in her field.

Theophilus Acheampong is an economist and risk analyst with over 14 years of experience in natural resource governance and public financial management. He has worked as an independent consultant on various global energy and mining industry projects, as well as public economics and market research assignments across frontier and emerging markets. His areas of specialisation include taxation and investment analysis—including extractive industries fiscal regime design and financial modelling—energy policy and international trade.

Acheampong's political risk expertise focuses on producing forward-looking reports and advising several global clients on sovereign risk, fiscal and monetary policy and regulatory developments in Sub-Saharan Africa. He holds a PhD in Economics and a Master of Science in Petroleum, Energy Economics and Finance from the University of Aberdeen, Scotland. He holds a Bachelor of Science in Chemical Engineering from the Kwame Nkrumah University of Science and Technology, Ghana. He is an associate lecturer at The Centre for Energy, Petroleum and Mineral Law and Policy (CEPMLP), University of Dundee, UK. He is also an associate lecturer and honorary research fellow at the Aberdeen Centre for Research in Energy Economics and Finance (ACREEF), University of Aberdeen, United Kingdom. Acheampong is a visiting fellow with the European Council on Foreign Relations (ECFR), working on the interface between energy transition, critical raw materials and international trade in Africa.

He is also a non-resident senior fellow at Ghanaian Think Tank IMANI Centre for Policy and Education.

Acheampong is quoted in several international and local media, including the BBC, Economist, WSJ, Bloomberg, AFP, CNBC and Al-Jazeera. He has over 70 research-based publications, including edited volumes, monographs, book chapters, refereed journal articles and commissioned reports in various fields. His recent co-edited book, *Petroleum Resource Management in Africa: Examining the Lessons from Ten Years of Oil and Gas Production in Ghana* (Palgrave Macmillan, 2022), has received wide acclaim.

List of Abbreviations

ASM	Artisanal and Small-Scale Mining
BIT	Bilateral Investment Treaty
CCDI	Climate Change Disclosure Index
CDM	Clean Development Mechanism
CDSB	Climate Disclosure Standards Board
CERD	Convention on Elimination of Racial Discrimination
CFU	Carbon Finance Unit
CIT	Corporate Income Tax
COMESA	Common Market for Eastern and Southern Africa
COP	Conference of the Parties
CRMs	Critical Minerals
DRC	Democratic Republic of the Congo
EEZ	Exclusive Economic Zones
EITI	Extractive Industries Transparency Initiative
ESG	Environmental, Social and Governance
ET	Emission Trading
EU	European Union
FDI	Foreign Direct Investment
FOCAC	Forum for China–Africa Cooperation
FPIC	Free, Prior and Informed Consent
GDP	Gross Domestic Product
GDPG	Gross Domestic Product Growth
GHG	Greenhouse Gases
GRI	Global Reporting Initiative
HDI	Human Development Index
ICCPR	International Covenant on Civil and Political Rights
ICJ	International Court of Justice

ICSID	International Centre for the Settlement of Investment Disputes
IFC	International Finance Corporation
IFRS	International Financial Reporting Standards
IGY	International Geophysical Year
IIRC	International Integrated Reporting Council
ILC	International Law Commission
INDC	Intended Nationally Determined Contribution
ISDS	Investor–State Dispute Settlement
LSM	Large-Scale Mining
MIGA	Multilateral Investment Guarantee Agency
MOP	Meeting of the Parties
NDCs	Nationally Determined Contributions
NO_x	Nitrogen Oxides
NYSE	New York Stock Exchange
ODS	Ozone-Depleting Substances
PRI	Principles of Responsible Investment
PSC	Production Sharing Contracts
SA	Republic of South Africa
SADC	Southern Africa Development Community
SDGs	Sustainable Development Goals
SDR	Sustainable Development Report
SO_x	Sulphur Oxides
TCFD	Task Force on Climate-related Financial Disclosures
UN	United Nations
UNECA	United Nations Economic Commission for Africa
UNEP	United Nations Environment Programme
UNFCC	United Nations Framework Convention on Climate Change
WHO	World Trade Organization
WSSD	World Summit on Sustainable Development

List of Figures

List of Tables

CHAPTER 1

Introduction: The Slow Evolution of the Global Mining Sector

Abstract This introductory chapter examines the gradual yet profound transformation of the global mining sector, situating its evolution within a broader historical and economic context. Minerals have long been foundational to human development, marking entire historical epochs such as the Bronze and Iron Ages and fuelling milestones like the Industrial Revolution. In the twenty-first century, their importance has grown, driven by urbanisation, industrialisation, population growth and rapid technological advancement.

Mineral raw materials, broadly classified into energy (e.g., coal, oil, gas) and non-energy types (e.g., industrial minerals like mica and talc, and construction materials like sand and granite), form the bedrock of various industries, from agriculture and construction to electronics and aviation. The sector has, however, undergone a strategic pivot in recent decades. While traditional resources like fossil fuels and base metals remain vital, attention has increasingly shifted towards critical minerals such as lithium, cobalt, nickel and rare earth elements. These materials are indispensable to the development of renewable energy systems, electric vehicles and high-tech electronics, positioning them at the core of the global energy transition.

This shift has redefined supply chains and global trade dynamics. Countries vary significantly in their capacity to access and supply critical minerals, creating geopolitical imbalances that influence renewable energy

© The Author(s), under exclusive license to Springer Nature Switzerland AG 2026
R. J. Heffron et al., *Revolution in the Global Critical Minerals Sector*, Just Transitions,
https://doi.org/10.1007/978-3-032-26366-7_1

development. For example, China's dominant position in critical mineral processing and trade has bolstered its influence, especially as international climate agreements like the 2015 Paris Agreement have accelerated demand for low-carbon technologies.

Despite the growing emphasis on critical minerals, traditional mineral resources continue to underpin global infrastructure and economic stability. Base metals remain crucial to construction, transportation and manufacturing, reinforcing the need for a balanced and inclusive perspective on mining sector priorities. This chapter sets the stage for a deeper exploration of the sector's structural changes, highlighting the complex interplay between legacy operations and emerging resource demands.

Keywords Global mining sector • Mineral resources • Critical minerals • Base metals • Energy transition • Renewable energy materials • Geopolitics of minerals • Supply chain dynamics • Industrial minerals • Low-carbon technologies • Economic history of mining

1.1 Introduction: The Global Mining Legacy[1]

Critical minerals are rising in value in how they will enable society to have a just sustainable climate transition. They form a critical part of the circular economy, the energy transition and are generally diffusing into the everyday life of society. Critical minerals being part of the mining sector have a long dark legacy of environmental damage and unjust financial distributions and this needs to change as they become central to the societal transitions needed to respond to the climate change challenges. In order to achieve this, this book explores the critical minerals industry and advances that an International Critical Minerals Treaty is needed, rather than fragmented bilateral treaties. It would benefit all countries and improve the position of society in achieving its climate and sustainable development goals.

The global climate change agenda has taken major steps forward since an instrumental treaty was signed in 2015 called the Paris Agreement. To move on to its next phase of evolution it needs to embrace its need for critical minerals as part of a sustainable and just economic transition. Therefore, to continue to deliver on climate change goals, it is advanced

[1] This subsection is with thanks also to the Journal of World Energy Law and Business for reproducing this in a short note (Heffron, RJ., 2026).

in this book that a new International Critical Minerals Treaty should be welcomed.

There have always been multiple problems with the mining sector that include accidents, major pollution, serious environmental impacts, unequitable financial distributions and carbon dioxide emissions. These have been documented in many countries and form an increasingly major source of climate risk.[2] Pressure is on governments and companies to be more transparent on critical minerals extraction due to their use and importance in the daily lives of citizens all over the world. Indeed, these resources are used in different forms in products and appliances that nearly all members of society interact with daily.

A key debate has emerged around how to improve the operation of the critical minerals sector over the last five years. Indeed, transitions research has highlighted the dark side of the mining sector and how it could play a greater role in resolving its problems.[3] Critical Minerals are also acknowledged in recent modelling as a key part of the circular economy and of the environment for sustainable societal transitions.[4] Today however, it is time to move the debate forward globally and in alignment with other international policy in particular on climate change and sustainability transitions. The mining sector has been beset by a legacy of unjust and unequitable outcomes, therefore, there is now a need to ensure that the full development of the critical minerals industry embraces a more 'just and equitable' approach. With the sector expected to grow by at least fourfold by 2040,[5] a new system is needed to govern this strategically important economic sector.

Research in academic literature has significantly advanced support for this approach through a diverse range of examples and empirical

[2] United Nations Environment Programme. 2024. Climate Risks in the Metals and Mining Sectors. Available at: https://www.unepfi.org/wordpress/wp-content/uploads/2024/05/Climate-Risks-in-the-Metals-and-Mining-Sector-1.pdf (last accessed 8 June 2025).

[3] Marin, A. and Gaya, D. 2021. Mining—The dark side of the energy transition. Environmental Innovation and Societal Transitions 41, 86–88.

[4] Lauer, A., de Castro, C. and Carpintero, O. 2025. Beyond Green capitalism: Global scenarios for fast societal transitions toward sustainability. Environmental Innovation and Societal Transitions 56, 100981.

[5] International Energy Agency. 2025. Global Critical Minerals Outlook 2025. Paris, France.

evidence.[6,7,8] In the policy area however, the United Nations have identified a clear path forward in this area with a recent explicit focus on critical minerals and justice and equity.[9] The UN report entitled *Resourcing the Energy Transition: Principles to Guide Critical Energy Transition Minerals towards Equity and Justice* is based around seven principles with actionable points for countries to follow. This is a significant development because the mining sector for years has struggled for any significant international policy action (despite the limited success of the Extractive Industries Transparency Initiative on oil, gas and coal revenues).

Notwithstamding recent research perspectives,[10] even the UN actions are narrow approaches, and not ambitious enough. Overall, what is needed is a new International Treaty for Critical Minerals. Significantly, there are early views from a number of countries, including Colombia and Brazil, who highlight that transparency and traceability are key for such a proposed Treaty.[11] These countries initially aimed to advance their version of such a Treaty at UN COP30 in 2025. There are clear benefits to such an International Treaty for Critical Minerals given the economic importance of these critical minerals. Supply and demand of these resources will impact economic stability into the future as it will affect a country's pathway to meeting climate, energy transition, sustainability and now, 2050 UN just transition goals and targets. However, critical minerals were deleted from the final text that was agreed upon in Brazil at UN COP30. To see this

[6] Ali, S. H., Giurco, D., Arndt, N., Nickless, E., Brown, G., Demetriades, A., Durrheim R., and others. 2017. "Mineral Supply for Sustainable Development Requires Resource Governance." Nature 543 (March 16): 367–72.

[7] Sovacool, B. K., Ali, S. H., Bazilian, M., Radley, B., Nemery, B., Okatz, J. and Mulvaney, D. 2020. Sustainable minerals and metals for a low-carbon future. Science, vol. 367 (6473) 30–33.

[8] Heffron, R. J. 2020. The Role of Justice in Developing Critical Minerals. The Extractives Industry and Society, 7 (3), 855–863.

[9] United Nations. 2024. Principles to Guide Critical Energy Transition Minerals towards Equity and Justice Available at: https://www.un.org/sites/un2.un.org/files/report_sg_panel_on_critical_energy_transition_minerals_11_sept_2024.pdf (last accessed, 8 June 2025).

[10] Ali. S. H. et al. 2025. A global minerals trust could prevent inefficient and inequitable protectionist policies. Science, 388 (6751) 1028–1030.

[11] Business and Human Rights Resource Centre. 2024. Colombia proposes a binding treaty for responsible mining of critical minerals at COP30Available at: https://www.business-humanrights.org/fr/derni%C3%A8res-actualit%C3%A9s/colombia-and-brazil-propose-a-binding-treaty-for-responsible-mining-of-critical-minerals-at-cop30/ (last accessed, 8 June 2025).

latter deletion as a negative is to misconstrue the issue that critical minerals are as closely important as the energy transition, the just transition and climate change.

A new global vision for mining has to envisage a critical minerals industry that delivers benefits (whether they be short, medium or long term) for all the key stakeholders such as governments, companies and civil society organisations. The aims must ensure that no one is left behind in the future development of the mining sector (i.e., align with the UN 'just transition pathway' approach under the UN Just Transition Work Programme 2024–2026). This approach over time has less risk for all stakeholders and can be accelerated by more collective action—such as through a global treaty.

An International Treaty for Critical Minerals should be governed through a clear and systematic Treaty grounded in principles of justice that delivers more just outcomes. Such a Treaty would be based on principles such as distributive, procedural, recognition, restorative and cosmopolitan justice. More will be discussed on this in Chap. 6. Too often, improvements in the mining sector have been limited, resulting in the perpetuation of the same legacy problems of increasingly contributing to climate risk.

Finally, there are different possible pathways forward for the mining sector with the advent of critical minerals, but none offers the transformational change that an International Treaty for Critical Minerals could deliver. In the years ahead, societies should not be looking at an industry that only aspired to create problems associated with past legacies of operations and practice. Instead, a new revolutionary approach to the industry is needed. The first steps to an International Treaty for Critical Minerals were, in essence, reflected in UN COP30 discussions. A key priority for the future will therefore be to establish critical minerals as a formal part of the international agenda. Critical minerals must be included in the debate on how societies can achieve enhanced climate mitigation and a sustainable societal transition. Indeed, these critical minerals and their use, play a strategic and significant role in determining whether the UN climate change and sustainability goals can be achieved.

1.2 An Overview and Background into the Global Mining Sector

Minerals are vital raw materials that support various economic sectors, playing a key role at the start of the value chain. Their demand has grown significantly in the twenty-first century due to urbanisation, industrialisation, population growth and technological progress.[12] Historically, minerals were central to the industrial revolution and defined entire eras like the Bronze and Iron Ages. Today, they are essential in industries such as automotive, aviation, electronics and agriculture.[13] Mineral raw materials are broadly categorised into energy (e.g., crude oil, natural gas, coal) and non-energy types, with the latter, including industrial minerals like talc and mica, and construction minerals like sand and granite being essential to many industries.[14]

Over recent decades, the global mining sector has undergone significant structural and strategic transformations. Once predominantly oriented towards the extraction of fossil fuels and industrial metals, the sector is now increasingly focused on what are termed *critical minerals*.[15] These minerals including, lithium, cobalt, nickel and various rare earth elements, are essential to the development of renewable energy technologies, electric vehicles and advanced electronics.[16] Their growing importance is directly tied to the global energy transition, as countries seek to decarbonise their economies and invest in sustainable infrastructure.

It is important to note that different countries vary in their ability to access and supply critical mineral resources, leading to differences in their roles within global trade networks, differences that significantly impact the development of renewable energy. For instance, China, as a major player in critical mineral trade, possesses strong trade capacity and central influence. These advantages support renewable energy growth, particularly in

[12] Zhu X, Ding Q, and Chen J, How does critical mineral trade pattern affect renewable energy development? The mediating role of renewable energy technological progress. Energy Economics. 2022 Aug 1;112:106164.

[13] Nalule VR, Mining and the Law in Africa: Exploring the social and environmental impacts. Springer Nature; 2019 Nov 22.

[14] Tiess, G., 2011. General and International Mineral Policy: Focus: Europe. New York: Springer Wien.

[15] Kalantzakos S. The race for critical minerals in an era of geopolitical realignments. The International Spectator. 2020 Jul 2;55(3):1–6.

[16] Hayes SM, McCullough EA. Critical minerals: A review of elemental trends in comprehensive criticality studies. Resources Policy. 2018 Dec 1;59:192–9.

high-income countries, and this positive effect has intensified since the adoption of the 2015 Paris Agreement and the global transition to a low carbon economy.[17]

While the twenty-first-century focus has increasingly shifted towards critical minerals, this trend has not diminished the importance of traditional mineral resources. Base metals remain fundamental to modern infrastructure and industrial development. It is therefore essential to maintain a balanced perspective that acknowledges both the growing emphasis on new resource frontiers and the enduring relevance of established extractive industries.

In line with this shift, many multinational mining companies are prioritising investments in critical minerals and scaling back operations in regions where gold and other non-critical metals are no longer economically viable. This withdrawal impacts national revenues while leaving behind abandoned mines that often become hotspots for illegal and artisanal small-scale mining (ASM) activities. For example, in January 2025, several illegal miners were trapped in South Africa while searching for gold; they became stranded during a nationwide police operation targeting illicit mining at disused sites remnants of an industry that once played a central role in the country's economy but has since declined.[18] As large firms exit, the lingering hope of finding leftover deposits continues to draw desperate individuals, especially undocumented migrants, into hazardous and unregulated mining activities.[19]

While large-scale mining (LSM) is often regarded as socially and environmentally safer than artisanal and small-scale mining (ASM), recent incidents highlight that significant risks still exist within LSM operations. For example, in May 2025, over 260 miners in South Africa were trapped underground following an accident at Sibanye-Stillwater's Kloof gold mine, one of the company's deepest mines, located approximately 60 km

[17] Zhu X, Ding Q, Chen J. How does critical mineral trade pattern affect renewable energy development? The mediating role of renewable energy technological progress. Energy Economics. 2022 Aug 1;112:106164.

[18] Mayeni Jones & Khanyisile Ngcobo, 'South Africa's Stilfontein Mine: The Blockaded Shaft Where Dozens Died' (BBC News, 18 January 2025) https://www.bbc.com/news/articles/c62qqg0zj6yo accessed 9 June 2025.

[19] Mayeni Jones & Khanyisile Ngcobo, 'South Africa's Stilfontein Mine: The Blockaded Shaft Where Dozens Died' (BBC News, 18 January 2025) https://www.bbc.com/news/articles/c62qqg0zj6yo accessed 9 June 2025.

west of Johannesburg.[20] The incident occurred when the hoist system used to access the shaft was damaged.[21] Despite extensive rescue efforts, the event underscores the persistent safety challenges faced by miners, regardless of whether they work in LSM or ASM, for both traditional minerals and critical minerals.

In light of the above, it is crucial to note that technological or market transformations in mining must be analysed in tandem with deeper ethical, environmental, social and political questions, particularly those relating to justice. Justice in the mining sector cannot be understood merely as legal compliance or the efficient allocation of revenues. It must be examined through the broader lens of distributive, procedural and environmental justice, each stage of the extractive lifecycle, from exploration and extraction to processing and closure, raises critical questions about who benefits, who bears the costs and who has a voice in decision-making processes.[22]

A key issue that arises in this context is the unequal application of justice across national and regional contexts. Developed countries such as Canada, the United States, Australia and Russia have largely succeeded in leveraging their mineral wealth for economic growth and national development, supported by relatively stable governance systems and institutional capacity.

Using Canada as an example, data[23] from the 2021 census demonstrates some key issues in their mining sector:

- It shows that over 17,300 Indigenous people were employed in the country's mining and mineral processing industries and approximately half of this employment was in the upstream mining subsector, where indigenous workers made up 11% of the labour force,

[20] Tim Cocks and Felix Njini, 'Sibanye workers begin to surface after accident at South African gold mine' (Reuters, 23 May 2025) https://www.reuters.com/sustainability/south-africas-mining-union-says-it-received-reports-300-trapped-mine-2025-05-23/ accessed 9 June 2025.

[21] Tim Cocks and Felix Njini, 'Sibanye workers begin to surface after accident at South African gold mine' (Reuters, 23 May 2025) https://www.reuters.com/sustainability/south-africas-mining-union-says-it-received-reports-300-trapped-mine-2025-05-23/ accessed 9 June 2025.

[22] Heffron RJ. Applying energy justice into the energy transition. Renewable and Sustainable Energy Reviews. 2022 Mar 1;156:111936.

[23] Natural Resources Canada, 'Minerals and the Economy' (Natural Resources Canada) https://natural-resources.canada.ca/minerals-mining/mining-data-statistics-analysis/minerals-economy. accessed 9 June 2025.

more than twice the national average across all industries, which stands at 4%.

- Revenues, in 2021, the mining sector and related support activities, including mining, quarrying and support services, contributed $5.4 billion in corporate income taxes and royalties. Of this total, 71% came from mining taxes and royalties, with the remaining portion paid as corporate income taxes to federal, provincial and territorial governments.
- In 2022, the minerals and metals sector directly contributed $109 billion to Canada's GDP, accounting for 6% of the national total. Indirect contributions added another $40 billion, bringing the sector's overall GDP impact to $149 billion.

This Canadian example demonstrates that with effective resource management, natural wealth can significantly boost a country's economy and enhance the financial well-being of its citizens. In contrast, many developing nations continue to grapple with challenges such as weak regulatory frameworks, limited local value addition, environmental degradation and inadequate community engagement.[24] Further, in many cases, mineral resources have contributed to political instability, social unrest and economic dependency.[25]

These disparities prompt a fundamental question: *How is justice conceptualised and practised in the mining sector, and how do those practices differ between developed and developing contexts?* This is not simply a theoretical concern; it is central to understanding the structural dynamics that shape global mining practices today. This chapter, and indeed the broader book, takes this question as a point of departure.

By interrogating the uneven benefits and burdens associated with mineral extraction, we aim to shed light on the systemic inequalities embedded in the global mining economy. Chapter Two will build on this foundation by offering a detailed examination of justice in mining, analysing both normative frameworks and practical case studies to consider how a more equitable and sustainable approach to resource governance might be realised.

[24] Nalule VR, editor. Mining Law and Governance in Africa: Transformation and Innovation for a Sustainable Mining Sector. Taylor & Francis; 2023 Jun 9.

[25] Okoi O, Nalule VR. Governing Natural Resources for Sustainable Peace in Africa. TAYLOR FRANCIS Limited; 2023.

Table 1.1 List of major initiatives on mineral development

Category 1	*Category 2*
• Millennium development goals (MDGs) • Sustainable development goals (SDGs) • OECD due diligence guidance for responsible Supply chains of minerals from conflict-affected and high-risk areas • Africa mining vision • Responsible mineral development initiative (RMDI) • Extractive industries transparency initiative (EITI)	• International finance corporation (IFC) Performance standards • Initiative for responsible mining assurance (IRMA) • Equator principles (EP) • International council on Mining & Metals (ICMM) • Responsible mining index (RMI) • United Nations global compact (UNGC) • Global reporting initiative (GRI) • Free, prior and informed consent (FPIC) dialogue • Conflict-free sourcing initiative (CFSI)/smelter program (CFSP) • Conflict-free tin initiative • Diamond development initiative (DDI) • Alliance for responsible mining (ARM) • Public-Private Alliance for responsible minerals trade

Source: Constructed by the Authors (2025)

Moving forward, we note that minerals and natural resources generally emerge at multiple levels—continental, regional, and national—shaped by the interplay of global narratives. These frameworks are designed to establish rules, institutions, and processes that promote transparent, accountable, and inclusive governance. A significant number of initiatives are underway regarding mineral development, with examples outlined in Table 1.1 and some of these are discussed in more detail in Chap. 5:

One of the key results from all these international initiatives must be their implementation. The growth and expansion of the mining sector will happen faster if these initiatives are applied according to specific key steps:

- The first is to create a stable investor framework guaranteeing legal certainty that ensures a return on their investment.

- A second result will be to ensure that the mining sector contributes overall to sustainable and just economic growth and avoids legacy problems as evidenced by past fossil fuel projects.
- A third result will be due to a lower risk premium on mining project development, and therefore enabling the project developer to secure (and repay) finance at lower cost. This will translate into the implementation of more mining projects, ensuring more taxation revenue, job creation and socio-economic impact.

In the next sub-section, we shall explore the different minerals and the different countries which are endowed with these minerals.

1.3 Mineral Resources by Country: An Overview

1.3.1 Critical Minerals

Critical minerals such as lithium, cobalt, nickel, and rare earth elements are vital to the global energy transition, powering electric vehicles, renewable energy systems, and energy storage technologies. As demand surges, these minerals have become central to global policy and trade, with supply security increasingly critical due to price volatility, supply chain disruptions and geopolitical risks.

Despite their crucial role in the twenty-first century, the distribution of critical minerals is highly uneven. Few countries hold significant reserves, and even fewer control refining capacity, creating a strategic vulnerability. Industrialised nations remain dependent on external suppliers, while producing countries often lack the infrastructure to process and add value.

According to the International Energy Agency (IEA), mining operations are expected to stay geographically concentrated. By 2035, the top three producers will supply: 85% of nickel (up from 75% in 2024), with similar concentration trends for copper and cobalt; Indonesia is set to see major growth in both nickel and cobalt output; and by 2035, China is expected to supply over 60% of refined lithium and cobalt, and around 80% of battery-grade graphite and rare earth elements.[26] These figures highlight the need to not only secure raw mineral supplies but also diversify processing capabilities to strengthen global supply chain resilience.

[26] IEA (2025), Global Critical Minerals Outlook 2025, IEA, Paris https://www.iea.org/reports/global-critical-minerals-outlook-2025, Licence: CC BY 4.0.

Data[27] demonstrates that critical minerals production is heavily concentrated in a few countries, raising significant geopolitical and ethical concerns. For example, China dominates global refining capacity, exerting major influence over supply chains, while many resource-rich but economically disadvantaged countries face challenges such as poor governance, environmental harm, and limited economic benefit. As such, critical minerals are not just industrial inputs—they are strategic assets that shape global power dynamics.

According to the IEA's recent data, demand is set to surge, primarily driven by the energy transition for example some of the key data issues the IEA highlights is:

- In the Stated Policies Scenario (STEPS), lithium demand is projected to grow fivefold by 2040; graphite and nickel will double;
- cobalt and rare earth elements will increase by 50–60%;
- copper demand is expected to rise by 30%
- these latter increases are largely due to the rise of electric vehicles, energy storage, grid electrification, and the growing need for permanent magnets;
- it is predicted that meeting this demand will require the major investment of USD 500 billion in mining by 2040 under STEPS, rising to USD 600 billion under the more ambitious Announced Pledges Scenario (APS).
- And these figures also reflect increasing project costs, especially in mature markets like copper, where ore quality is declining.[28]

To ensure secure and equitable access, coordinated global action is essential to diversifying supply chains, investing in responsible extraction, supporting local value addition, and advancing recycling and substitution. A global governance framework that addresses both supply security and distributive justice is crucial. In this context, critical minerals are becoming key to the geopolitical dynamics of the energy transition, shaping debates on sustainability, equity and long-term economic resilience. In the next

[27] N E Idoine and others, *World Mineral Production 2019–23* (British Geological Survey 2025) https://nora.nerc.ac.uk/id/eprint/539285/ accessed 9 June 2025.

[28] IEA (2025), Global Critical Minerals Outlook 2025, IEA, Paris https://www.iea.org/reports/global-critical-minerals-outlook-2025, Licence: CC BY 4.0.

sub-section, we briefly discuss the nexus between critical minerals and security.

1.3.2 Nexus Between Critical Minerals, Peace, and Security

While minerals have long been associated with conflict and insecurity, often referred to as 'conflict minerals' in regions such as Central Africa and the DRC, recent developments illustrate their potential to serve as strategic tools for peacebuilding and post-conflict recovery. Increasingly, states are leveraging these resources beyond economic development to foster geopolitical alliances and stabilise fragile contexts.

A prominent and very recent example is the May 2025 minerals agreement between Ukraine and the United States of America, which was formalised in Washington, D.C. The deal establishes the United States–Ukraine Reinvestment Fund, a jointly managed investment mechanism aimed at supporting Ukraine's reconstruction following years of conflict.[29] Under this agreement, the U.S. receives preferential access to Ukraine's critical mineral reserves, including rare earth elements, titanium and lithium, in exchange for financial contributions and potential military assistance, including air defence systems.[30] Ukraine, in turn, commits 50% of future revenues from new resource licences to the fund.[31] Data from Ukraine's Ministry of Economy indicate that the country holds deposits of 22 of the 34 minerals classified as critical by the European Union. These include titanium, zirconium, graphite, lithium and various precious and non-ferrous metals. Ukraine also possesses significant reserves of rare earth elements (REEs) such as lanthanum, cerium and neodymium, which are essential for high-tech industries, including electronics, defence, aerospace and renewable energy.[32] The country accounts for approximately 7% of

[29] Al Jazeera, 'Ukraine's parliament ratifies landmark minerals deal with US' (Al Jazeera, 8 May 2025) https://www.aljazeera.com/news/2025/5/8/ukraines-parliament-ratifies-landmark-minerals-deal-with-us accessed 23 May 2025.

[30] Al Jazeera, 'Ukraine's parliament ratifies landmark minerals deal with US' (Al Jazeera, 8 May 2025) https://www.aljazeera.com/news/2025/5/8/ukraines-parliament-ratifies-landmark-minerals-deal-with-us accessed 23 May 2025.

[31] Al Jazeera, 'What Is in the US-Ukraine Minerals Deal?' (Al Jazeera, 1 May 2025) https://www.aljazeera.com/news/2025/5/1/what-is-in-the-us-ukraine-minerals-deal accessed 23 July 2025.

[32] BBC News, *US and Ukraine sign critical minerals deal* (BBC News, 30 April 2025) https://www.bbc.com/news/articles/c5ypw7pn9q3o accessed 1 July 2025.

global titanium production, and its largely untapped lithium reserves, estimated at 500,000 tonnes, are among the largest in Europe[33] and probably why it is an issue that has been brought to prominence over the last few years in the contract of the Russia-Ukraine war.

Beyond its economic implications, this Ukraine-USA deal carries important geopolitical significance. It signals U.S. commitment not only to Ukraine's recovery but also to its long-term sovereignty and stability. The agreement may also be interpreted as a diplomatic message opposing the ongoing war with Russia, a conflict that began in 2014 with the annexation of Crimea and escalated dramatically with Russia's full-scale invasion in February 2022. While earlier peace efforts have largely failed, such strategic partnerships could offer new pathways for conflict resolution and deterrence.[34]

This use of critical minerals as instruments of diplomacy marks a shift from their more traditional association with instability. In regions such as the Democratic Republic of Congo (DRC), mineral wealth has historically fuelled violence. Armed groups have exploited resources like coltan, gold and tin to finance their operations, prolonging insecurity and undermining state authority—indeed conflict and critical minerals is already a growing area of research. The ongoing conflict in the Democratic Republic of the Congo (DRC), particularly in its mineral-rich eastern provinces, has long illustrated the darker side of global demand for critical minerals. While the roots of the conflict are complex, encompassing historical, ethnic and political dimensions, control over the DRC's vast reserves of high-value resources such as cobalt, coltan, tin, tantalum and gold has consistently played a central role in fuelling violence and instability. Armed groups, including the M23 movement, have vied for control of mining areas, often with the involvement or backing of regional actors like Rwanda and Uganda. These minerals, essential to global supply chains for electronics, batteries and renewable energy technologies, have paradoxically contributed to prolonged insecurity and humanitarian crises in the region.

However, a significant shift occurred on 27 June 2025, when the United States brokered a peace agreement between the DRC and Rwanda in Washington, D.C. This accord marks a transformative approach to

[33] BBC News, *US and Ukraine sign critical minerals deal* (BBC News, 30 April 2025) https://www.bbc.com/news/articles/c5ypw7pn9q3o accessed 1 July 2025.

[34] BBC News, 'US and Ukraine sign critical minerals deal' (BBC News, 30 April 2025) https://www.bbc.com/news/articles/c5ypw7pn9q3o accessed 20 May 2025.

conflict resolution, wherein access to critical mineral resources is directly linked to commitments for regional stabilisation and security.[35] Under the terms of the agreement, Rwanda pledged to withdraw its troops from eastern DRC within 90 days, while the Congolese government agreed to dismantle support for the Democratic Forces for the Liberation of Rwanda (FDLR) militia.[36] Both nations committed to establishing a joint security coordination mechanism and a regional economic integration framework, with the United States facilitating investments in infrastructure, mining and hydropower projects.

This agreement exemplifies a novel 'minerals-for-peace' model, where resource governance is intertwined with peacebuilding efforts. By aligning economic incentives with security objectives, the accord aims to transform mineral wealth from a source of conflict into a foundation for sustainable development and regional cooperation. The involvement of the United States underscores the strategic importance of critical minerals in global geopolitics, particularly in countering external influences and ensuring secure supply chains for emerging technologies.

In contrast to past scenarios where mineral wealth exacerbated tensions, this initiative demonstrates the potential for critical resources to serve as instruments of peace and stability. It underscores the imperative for innovative approaches that integrate resource management with comprehensive peace strategies, ensuring that mineral-rich regions can transition from conflict to prosperity.

Similarly, in Nigeria's Niger Delta, oil extraction has led to widespread environmental degradation, marginalisation of local communities, and insurgent activity by groups such as the Movement for the Emancipation of the Niger Delta (MEND). Yet, the narrative is not universally bleak. Examples such as Botswana demonstrate that with strong institutions, transparent governance and equitable revenue distribution, resource wealth can be a foundation for stability and development. Botswana's effective management of its diamond sector stands in stark contrast to the chaos seen in less regulated contexts, highlighting the critical role of

[35] Al Jazeera, 'DR Congo and Rwanda sign peace deal in turning point after years of war' (Al Jazeera, 27 June 2025) https://www.aljazeera.com/news/2025/6/27/dr-congo-and-rwanda-sign-peace-deal-in-turning-point-after-years-of-war accessed 2 July 2025.

[36] Al Jazeera, 'DR Congo and Rwanda sign peace deal in turning point after years of war' (Al Jazeera, 27 June 2025) https://www.aljazeera.com/news/2025/6/27/dr-congo-and-rwanda-sign-peace-deal-in-turning-point-after-years-of-war accessed 2 July 2025.

governance in determining whether minerals become a curse or a catalyst for peace.

In sum, while critical minerals carry inherent risks, particularly in fragile or poorly governed settings, they also present new opportunities for diplomacy, reconstruction, and international cooperation. The evolving role of these resources underscores the need for policies that integrate security, governance and economic development to ensure that mineral wealth contributes to long-term peace and global stability.

1.4 Conclusion and Scope of the Text

This book offers a detailed exploration of the global mining sector's evolution and its critical role in today's extractive and energy landscapes. It opens with an introduction tracing the slow but significant development of the mining industry, setting the historical and structural context for subsequent analysis. The text then delves into justice pathways within extractives and critical minerals, emphasising frameworks for equitable resource governance, community rights, and social inclusion. Then there is a focus on economic factors with case studies on Africa and it explores the shaping of the global mining industry alongside the geopolitical dynamics surrounding critical minerals, which are essential to modern technologies and the energy transition. Sustainability challenges and strategies are considered to ensure mining contributes positively to environmental goals. The final chapters present future outlooks and recommendations aimed at guiding policy and industry towards a more just, sustainable and resilient mining future.

The global minerals landscape is defined by diverse geological endowments, uneven access, and shifting priorities shaped by the energy transition, sustainability concerns and geopolitical realities. While Europe has long benefited from a well-documented and state-managed extractive tradition, it now grapples with declining energy mineral production and increasing reliance on imports for strategic metals. Nevertheless, its domestic capacity for industrial and construction minerals, combined with robust institutional oversight, provides a strong foundation for pursuing critical mineral security in the context of the green and digital transitions.[37]

[37] N E Idoine and others, *World Mineral Production 2019–23* (British Geological Survey 2025) https://nora.nerc.ac.uk/id/eprint/539285/ accessed 9 June 2025.

Africa's mineral wealth, holding roughly 30% of the world's reserves, positions it as a key player in global supply chains, particularly for cobalt, platinum group metals and emerging lithium sources. Yet, historical patterns of extraction, weak governance and socio-environmental challenges continue to limit the continent's ability to fully leverage this potential for inclusive and sustainable development. Efforts like the African Mining Vision signal a growing commitment to transparency, regional cooperation and value addition, though effective implementation remains critical.

In Asia and the Americas, mineral strategies are increasingly driven by national policies aimed at securing supply chains, encouraging domestic value addition, and addressing sustainability concerns. China and Indonesia have emerged as central to global critical mineral processing, while the United States, Canada and key Latin American countries are responding with targeted legislation and sustainable mining initiatives. Across both regions, however, tensions between development, environmental preservation and community rights persist.

Globally, the evolution of international frameworks such as the Extractive Industries Transparency Initiative (EITI) and the Kimberley Process highlights the importance of governance in ensuring that resource wealth translates into equitable development. Region-specific policies like South Africa's Mining Charter exemplify efforts to link mineral production with social inclusion, though challenges in enforcement and regulatory clarity remain.

Together, these regional experiences underscore a shared imperative: to balance mineral development with accountability, equity and long-term resilience. As demand for minerals, especially those critical to the energy transition, accelerates, the need for inclusive, transparent and responsible resource governance becomes more urgent than ever. This chapter sets the stage for a deeper analysis of how different countries and regions are navigating this complex terrain. The next chapters will delve into the topic of justice in the mining sector and the economic impact of mining.

CHAPTER 2

Justice Pathways in Extractives and Critical Minerals

Abstract Critical mineral use is on the rise in daily life. One key aspect that has arisen is how these minerals will be developed in a just way. This has been an increasing focus of researchers over the last decade. However, this is a topic that is now a key policy concern as the United Nations in 2024 has directly published on how greater justice and equity can happen in this growth area of critical minerals. This chapter and this book provide a deeper and more holistic view into justice and critical mineral extraction and production. This conceptual chapter makes an original contribution that utilises an interdisciplinary perspective, involving legal geography, and explores key issues of justice that include distributive, procedural, restorative, recognition and cosmopolitan. It identifies the key questions that need to be resolved under each element of justice and the unfortunate, limited timeframes for action. 'Critical justice' areas include taxation, environmental impact assessments, waste management, social licence to operate and cross-border actions. Resolving these issues will directly address societal issues of inequality and ensure a just transition to a low-carbon economy. Already, there is a global race for critical minerals, and justice needs a stronger role in its development based on evidence to date. Finally, this chapter details the JUST Framework, which adds a more systematic and real-world perspective to the issues, as it also analyses justice issues within a time frame and their geographical scope.

© The Author(s), under exclusive license to Springer Nature Switzerland AG 2026
R. J. Heffron et al., *Revolution in the Global Critical Minerals Sector*, Just Transitions,
https://doi.org/10.1007/978-3-032-26366-7_2

Keywords Critical minerals • Mineral extraction • Resource justice • Environmental justice • Social licence to operate • JUST Framework • Low-carbon transition • Distributive and procedural justice • Sustainable mining • Cross-border governance • Legal geography

2.1 Introduction

Critical minerals have long been essential for society, but they are now considered super-critical. Society has always faced resource challenges, which is why economics remains vital, as it fundamentally deals with managing resources efficiently. Today, however, society must address the acknowledged impacts of climate change and (through international legal agreements such as the 2015 Paris COP21 Agreement) has committed to developing a low-carbon economy. Some countries have even set more ambitious targets, aiming for zero-carbon economies or aspiring to such goals, like the European Union's 2050 objective. These low-carbon ambitions, however, impose new demands on resources, making critical minerals even more essential.

The number and definition of critical minerals have changed over time, and this will probably remain the case as technological development increases to meet the demands of a low-carbon society. There is little literature in the area of critical minerals from a non-science perspective, but the non-science issues have been raised very recently in prominent journals such as *Nature*[1] and *Science.*[2]

Regardless, what is really missing from existing analysis is a perspective of justice as more and more of these critical minerals are extracted and used in society. This follows some recent research in terms of the extractives industry that has already begun to highlight justice issues (in particular, distributive justice) from a taxation perspective on critical mineral extraction in Africa.[3] Further research has been conducted broadly on

[1] Ali Saleem H., Damien Giurco, Nicholas Arndt, Edmund Nickless, Graham Brown, Alecos Demetriades, Ray Durrheim, and others. 2017. "Mineral Supply for Sustainable Development Requires Resource Governance." Nature 543 (March 16): 367–72.

[2] Benjamin K. Sovacool, Saleem H. Ali, Morgan Bazilian, Ben Radley, Benoit Nemery, Julia Okatz, and Dustin Mulvaney. 2020. Sustainable minerals and metals for a low-carbon future. Science, vol. 367 (6473) 30–33.

[3] Heffron, R. J. 2018. The application of distributive justice to energy taxation utilising sovereign wealth funds. Energy Policy, 122, 649–654.

critical minerals[4] but the justice perspective lacks a concrete principle-based approach. However, most recently, the United Nations worked on the issue of justice, and produced the *Principles to Guide Critical Energy Transition Minerals Towards Equity and Justice*.[5] This is a significant policy document, which is why there is a need to focus on this issue more broadly.

This chapter aims to explore, from a conceptual and comparative perspective, the core justice aspects of critical mineral development and extraction. It will examine the issue from an interdisciplinary perspective utilising legal, economic and business taxation perspectives to deliver on what the key challenges for achieving justice in terms of critical minerals are. This is necessary considering their increased use for the development of a low-carbon economy. The underlying focus here is to ensure that the transition to a low-carbon economy happens in a 'just' way,[6] and for that to be achieved, all the elements of justice need to be addressed, such as distributive, procedural, recognition, restorative and cosmopolitanism.

To ascertain what the key justice challenges for critical mineral development and extraction are, this chapter first looks briefly at the critical minerals industry (Section 2.2). Then, a literature review on critical minerals and justice is analysed (Section 2.3), and the need for justice in this sector is explored in Section 2.4. Following that, a conceptual review of existing justice issues is presented in Section 2.5, utilising the *JUST* framework from law and geography. Section 2.6 then advances conclusions on justice and critical minerals and identifies the pathways for future research. The aforementioned legal geography analysis makes an original contribution to this new and growing area of research on critical minerals, where, as stated earlier, much of the research lies in science.

[4] Nakanwagi, S. 2024. Critical Minerals, Sustainability and the Energy Transition in the Global South: A Justice Perspective. Bloomsbury.

[5] United Nations. 2024. Principles to Guide Critical Energy Transition Minerals towards Equity and Justice Available at: https://www.un.org/sites/un2.un.org/files/report_sg_panel_on_critical_energy_transition_minerals_11_sept_2024.pdf (last accessed, 8 June 2025).

[6] Heffron R. J. and McCauley, D. 2018. What is the 'Just Transition'? Geoforum, 88, 74–77.

2.2 Overview of Critical Minerals Industry

As stated earlier, critical minerals have long been part of the global economy. However, the surge in demand driven by the low-carbon energy transition is responsible for their return to prominence. Though the US already commissioned a first report on their strategy for critical minerals due to the clean energy technology back in 2010,[7] this followed an earlier report from 1973,[8] building on the Strategic and Critical Minerals Stock Piling Act of 1939 (P.L. 96–41, 1939).

The 1973 report emphasised the importance of these minerals by posing three fundamental: "*(1) How important is it to our present industrial civilization and standard of living? (2) how much of it do we have and to what extent is it economically and technologically available? and (3) how and where can we find more?*"[9] There are different definitions of what 'critical' is. The Strategic and Critical Minerals Stock Piling Act of 2005 (50 U. S. C. 98 et seq.) states that strategic and critical minerals are those that are needed for military, industry and civilian use. While the legislation does not differentiate between strategic and critical, the latter would be broader in that strategic is more military related,[10] i.e., a mineral would be more often critical than strategic.[11] Today, in the context of the energy transition, one of the most clear and distinct definitions for critical minerals is that 'critical' refers to their '*importance to the clean energy economy and risk of supply disruption*'.[12] The remainder of this chapter examines what is included in the critical minerals industry before detailing its value today.

[7] US Department of Energy (US, DOE, 2010). 2010. Critical Materials Strategy. Available at: https://www.energy.gov/sites/prod/files/edg/news/documents/criticalmaterialsstrategy.pdf (last accessed, 8 June 2025).

[8] U.S. Geological Survey (US GS). 1973. United States Mineral Resources. Professional Paper 820, (Brobst, D. A. and Pratt, W. P. (eds.)). Available at: https://pubs.er.usgs.gov/publication/pp820, (last accessed, 8 June 2025).

[9] U.S. Geological Survey (US GS). 1973: 1. United States Mineral Resources. Professional Paper 820, (Brobst, D. A. and Pratt, W. P. (eds.)). Available at: https://pubs.er.usgs.gov/publication/pp820, (last accessed, 8 June 2025).

[10] National Research Council. 2008. Minerals, Critical Minerals and the US Economy. National Academies Press: Washington DC, US.

[11] Robinson, A. L. 1986. Congress critical of foot-dragging on critical materials. Science, 234 (4772) 20–21.

[12] US Department of Energy (US, DOE, 2010). 2010: 6. Critical Materials Strategy. Available at: https://www.energy.gov/sites/prod/files/edg/news/documents/criticalmaterialsstrategy.pdf (last accessed, 8 June 2025).

2.2.1 What Is Included in the Critical Minerals Industry?

The second aspect of understanding what critical minerals are is what is included in the critical minerals (CRM) industry. This chapter does not seek to examine the industry's scope in detail; rather, it provides an overview of what is included before considering how, and to what extent, justice has been incorporated into the industry. Nevertheless, concerning the CRM, this paper takes a broad definition and includes elements and metals, energy mineral resources and supply chain products—by utilising the term CRM, it is recognised that the industry involves the raw material product, i.e. the modified mineral or element. The number of minerals that are classed as critical (according to the definition provided earlier—i.e. that critical refers to the importance to the clean energy economy and at risk of supply disruption) is increasing, and a clear example of this is in the EU, which classifies them as critical raw materials (CRMs). The EU identified 14 CRMs in 2011,[13] expanded the list to 20 in 2014,[14] and increased it further to 27 CRMs in 2017.[15] The European Commission classified CRMs according to economic importance and supply risk,[16] similar to the earlier-mentioned US definition above.

A review research study that analysed the number of critical minerals identified by reports from across the world held that the following were the most critical minerals: Rare earth elements (REE), Platinum Group Metals (PGMs), Indium, Tungsten, Germanium, Cobalt, Niobium, Tantalum, Gallium, Antimony, Bismuth, Thallium and Magnesium.[17] The latter review, however, recognised that this list of critical minerals would continue to expand. In that context, this chapter draws on that key

[13] European Commission. 2011. Tackling the Challenges in Commodity Markets and on Raw Materials. Available at: https://eur-lex.europa.eu/legal-content/EN/TXT/?uri=CELEX:52011DC0025 (Last accessed, 10 April 2020).

[14] European Commission. 2014. On the review of the list of critical raw materials for the EU and the implementation of the Raw Materials Initiative. Available at: https://eur-lex.europa.eu/legal-content/EN/TXT/?uri=CELEX:52014DC0297 (last accessed, 8 June 2025).

[15] European Commission. 2017. 2017 list of Critical Raw Materials for the EU. Available at: https://eur-lex.europa.eu/legal-content/EN/TXT/?uri=CELEX:52017DC0490 (last accessed, 8 June 2025).

[16] European Commission. 2011. Tackling the Challenges in Commodity Markets and on Raw Materials. Available at: https://eur-lex.europa.eu/legal-content/EN/TXT/?uri=CELEX:52011DC0025 (last accessed, 8 June 2025).

[17] Hayes, S. M. and McCullough, E. A. 2018. Critical minerals: A review of elemental trends in comprehensive criticality studies. Resources Policy, 59, 192–199.

Table 2.1 Summary of what is included in the critical minerals industry 2020

Summary of critical minerals 2020			
Aluminium	*Fluorspar*	*Manganese*	*Tellurium*
Antimony	Gallium	Niobium	Tin
Arsenic	Germanium	Platinum Group Metals (PGMs)	Titanium
Baryte	Natural graphite	Potash	Tungsten
Beryllium	Hafnium	Rare Earth Elements (REE)	Uranium
Bismuth	Helium	Rhenium	Vanadium
Caesium and Rubidium	Indium	Scandium	Zirconium & Hafnium
Chromium	Lithium	Strontium	
Cobalt	Magnesium	Tantalum	

Source: Created by the Author (2020)

research article[18] and adopts the EU and US perspective that the number of critical minerals would continue to grow as demand accelerates and technological development advances. Accordingly, this book adopts the US classification of critical minerals as identified in Table 2.1,[19] and this list will be expected to grow as the energy transition gathers pace.

2.2.2 The Value of the Critical Minerals Industry

It is no surprise that there is a range of different values placed on the critical minerals industry. Indeed, few reports seem willing to place a value on this growing industry, and they all agree it is growing and will grow nearly exponentially.[20] In addition, it is a global industry as the EU has

[18] Hayes, S. M. and McCullough, E. A. 2018. Critical minerals: A review of elemental trends in comprehensive criticality studies. Resources Policy, 59, 192–199.

[19] Fortier, S.M., Nassar, N.T., Lederer, G.W., Brainard, J., Gambogi, J., McCullough, E.A., 2018. Draft critical mineral list—Summary of methodology and background information—U.S. Geological Survey Technical Input Document in Response to Secretarial Order No. 3359: U.S. Geological Survey Open-File Report 2018–1021. https://doi.org/10.3133/ofr20181021.(15 p.).

[20] Benjamin K. Sovacool, Saleem H. Ali, Morgan Bazilian, Ben Radley, Benoit Nemery, Julia Okatz, and Dustin Mulvaney. 2020. Sustainable minerals and metals for a low-carbon future. Science, vol. 367 (6473) 30–33.

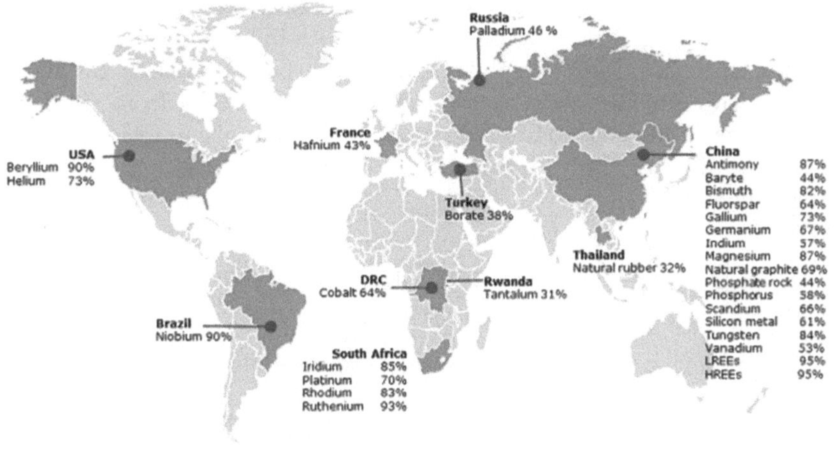

Fig. 2.1 Countries accounting for the largest share of global supply of CRMS. (Source: European Commission (2018))

highlighted in their report, depicted by Fig. 2.1[21] as well as the distribution of large shares of CRMS across the world.

The mining sector declined over the period 2011–2015 however, it has rebounded since then.[22] However, there remain expected issues according to many of the commercial practitioner reports, and in particular, they are environmental, social, and governance (ESG) issues and also a need to invest circa $240bn according to some reports.[23] These mineral resources play a dominant role in 81 countries that collectively account for a quarter of world GDP, half of the world's population and nearly 70% of those in extreme poverty.[24] Recent British Geological Survey (BGS) (2020) data

[21] European Commission 2018. Report on Critical Raw Materials and the Circular Economy. Available at: https://op.europa.eu/en/publication-detail/-/publication/d1be1b43-e18f-11e8-b690-01aa75ed71a1/language-en/format-PDF/source-80004733 (last accessed, 8 June 2025).

[22] British Geological Survey. 2020. World Mineral Production 2014–2018. British Geological Survey, Keyworth, Nottingham.

[23] Financial Times (Kettle, J. 19 February), 2020. Mining sector faces energy transition conundrum. Available at: https://www.ft.com/content/36ef7ab2-4f44-11ea-95a0-43d18ec715f5 (last accessed, 8 June 2025).

[24] World Bank. 2017. The Growing Role of Minerals and Metals for a Low Carbon Future. Available at: http://documents.worldbank.org/curated/en/207371500386458722/pdf/117581-WP-P159838-PUBLIC-ClimateSmartMiningJuly.pdf(last accessed, 8 June 2025).

highlights the acute demand for some resources. In particular, between 2017 and 2018, the BGS highlights the growth in lithium (an increase of 25%), cobalt (up 18%), graphite and nickel (both increased by about 13%).[25] Lithium has risen considerably, rising to 235% from 2014 to 2018 alone.[26] Indeed, the OECD stated that the use of extracted key metals will double by 2060, irrespective of technology change or efficiency improvements.[27]

2.3 Justice within the Critical Minerals Sector: A Literature Review

It is important to remember that the critical minerals sector forms part of the broader mining sector, which the UN has identified as crucial to achieving the Sustainable Development Goals[28]—the mining sector includes mineral exploration, extraction, milling/concentrating, smelting, refining, processing, transport and sale. Nevertheless, there is little literature, either research or practitioner-focused, which addresses issues of justice in the development and extraction of critical minerals. Predominantly, practitioner and policy literature focuses, in essence, on; (1) developing the critical minerals business, (2) rising demand, (3) raising finance and (4) political issues.[29] There is one EU policy report, however, that

[25] European Commission 2018. Report on Critical Raw Materials and the Circular Economy. Available at: https://op.europa.eu/en/publication-detail/-/publication/d1be1b43-e18f-11e8-b690-01aa75ed71a1/language-en/format-PDF/source-80004733 (last accessed, 8 June 2025).

[26] British Geological Survey. 2020. World Mineral Production 2014–2018. British Geological Survey, Keyworth, Nottingham.

[27] OECD. 2019. Global Material Resources Outlook to 2060: Economic Drivers and Environmental Consequences, OECD Publishing, Paris, France. Available at: https://doi.org/10.1787/9789264307452-en (last accessed, 8 June 2025).

[28] United Nations (UN). 2016. Mapping Mining to the SDGs: An Atlas. Available at: https://www.undp.org/content/undp/en/home/librarypage/poverty-reduction/mapping-mining-to-the-sdgs%2D%2Dan-atlas.html(Last accessed, 10 April 2020).

[29] This a short analysis from the following key articles: (1) Akong, C. 2020. Reframing matter: Towards a material discursive framework for Africa's minerals. The Extractive Industries and Society, 7 (2) 461–469; (2) Andersson, P. 2020. Chinese assessments of "critical" and "strategic" raw materials: Concepts, categories, policies, and implications. The Extractive Industries and Society, 7 (1) 127–137; (3) Bazilian, M. D. 2018. The mineral foundation of the energy transition. The Extractive Industries and Society, 5 (1) 93–97; (4) Coulomb, R., Dietz, S., Godunova, M. and Bligaard Nielsen, T. 2015. Critical minerals today and in 2030: an analysis of OECD countries. Prepared for the OECD by (Grantham

highlights 'fairness;' in effect, it frames fairness in terms of ensuring the EU's access to these critical raw materials at fair prices.[30] This represents a very limited form of justice that appears to prioritise large businesses and industry in general. This EU approach continues today whereby justice lacks mention in recent EU literature on legislating for critical raw minerals.

There is existing literature aimed at reviewing some considerations of justice, although actual references to justice within the text are rare.[31] Meanwhile, other researchers have assessed justice issues in the context of a single metal such as cobalt and a focus on a single justice issue, for example, distributive justice.[32] Further, while a UN (2019) report narrowly focuses its attention on the role of law, it is mainly in the context of procedural justice as a general focus in the mining sector.[33] That UN report was preceded by another general mining report in 2018 that focused on access to justice by highlighting procedural and, to a certain degree,

Research Institute on Climate Change and the Environment at the London School of Economics and Political Science (LSE), London, U.K.). Available at: http://www.lse.ac.uk/GranthamInstitute/wp-content/uploads/2015/11/OECD_minerals_paper_Coulomb_et al.pdf (last accessed, 8 June 2025); (5) He, Y. 2018. The trade-security nexus and U.S. policy making in critical minerals. Resources Policy, 59, 238–249; (6) Lee, J., Bazilian, M., Sovacool, B., Hund, K., Jowitt, S. M., Nguyen, T. P., Månberger, A., Kah, M., Greene, S., Galeazzi, C., Awuah-Offei, K., Moats, M., Tilton, J. and Kukoda, S. 2020. Reviewing the material and metal security of low-carbon energy transitions. Renewable and Sustainable Energy Reviews, 124, 109,789; (7) Viebahn, P., Soukup, O., Samadi, S., Teubler, J., Wiesen, K., Ritthoff, M. 2015. Assessing the need for critical minerals to shift the German energy system towards a high proportion of renewables. Renewable and Sustainable Energy Reviews, 49, 655–671: and (8) Benjamin K. Sovacool, Saleem H. Ali, Morgan Bazilian, Ben Radley, Benoit Nemery, Julia Okatz, and Dustin Mulvaney. 2020. Sustainable minerals and metals for a low-carbon future. Science, vol. 367 (6473) 30–33.

30 European Commission. 2008. The raw materials initiative — meeting our critical needs for growth and jobs in Europe. COM (2008) 699. Available at: https://eur-lex.europa.eu/legal-content/EN/TXT/PDF/?uri=CELEX:52008DC0699&from=EN (last accessed, 8 June 2025).

31 McCellan, B.C., Corder, G. D., and Ali, S. H. 2013. Sustainability of Rare Earths—An Overview of the State of Knowledge Minerals 2013, 3, 304–317.

32 Heffron, R. J. 2018. The application of distributive justice to energy taxation utilising sovereign wealth funds. Energy Policy, 122, 649–654.

33 United Nations (UN). 2019. Assessing the Rule of Law in Public Administration: The Mining Sector. Available at: https://www.undp.org/content/undp/en/home/librarypage/poverty-reduction/environmental-governance/ROL-MiningSector.html (last accessed, 8 June 2025).

recognition justice.[34] In contrast, this chapter aims to be far more comprehensive in its focus on justice, recognising that the development of critical minerals is a global issue, they are part of global business value and supply chains and they play a significant role in ensuring there is a just transition to a low-carbon economy across the world.

There has been a growth in literature on critical minerals over the last few years and this will increase. Despite this, issues of justice for this sector remain largely unexplored. Research on justice in the critical minerals sector should focus on the critical mineral development while reflecting their role in the transition to a low-carbon economy. As stated by Ali et al. in *Nature* in 2017, "*a transition to a low carbon society, [is] a change that will require vast amounts of metals and minerals. Mineral resourcing and climate change are inextricably linked, not only because mining requires a large amount of energy, but also because the world cannot tackle climate change without adequate supply of raw materials to manufacture clean technologies*".[35] Earlier and more recent influential studies have been published in leading journals such as *Nature Geoscience*,[36] *Science*[37] and *Renewable and Sustainable Energy Reviews*,[38] however, they lack a direct focus on justice issues. Environmental and social issues are recognised in the existing literature but receive only limited attention, and this chapter seeks to address that gap by examining how issues of justice (including fairness and equity) can be understood and resolved within the critical minerals sector.

[34] United Nations (UN). 2018. A Guide for Governments and Partners to Integrate Environment and Human Rights into the Governance of the Mining Sector. Available at: https://www.undp.org/content/undp/en/home/librarypage/poverty-reduction/environmental-governance/extracting-good-practices%2D%2Da-guide-for-governments-and-partners-.html (last accessed, 8 June 2025).

[35] P. 367, Ali Saleem H., Damien Giurco, Nicholas Arndt, Edmund Nickless, Graham Brown, Alecos Demetriades, Ray Durrheim, and others. 2017. "Mineral Supply for Sustainable Development Requires Resource Governance." Nature 543 (March 16): 367–72.

[36] Vidal, Olivier, Bruno Goffé, and Nicholas Arndt. 2013. Metals for a Low-Carbon Society. Nature Geoscience 6 (October): 894–96.

[37] Benjamin K. Sovacool, Saleem H. Ali, Morgan Bazilian, Ben Radley, Benoit Nemery, Julia Okatz, and Dustin Mulvaney. 2020. Sustainable minerals and metals for a low-carbon future. Science, vol. 367 (6473) 30–33.

[38] Lee, J., Bazilian, M., Sovacool, B., Hund, K., Jowitt, S. M., Nguyen, T. P., Månberger, A., Kah, M., Greene, S., Galeazzi, C., Awuah-Offei, K., Moats, M., Tilton, J. and Kukoda, S. 2020. Reviewing the material and metal security of low-carbon energy transitions. Renewable and Sustainable Energy Reviews, 124, 109789.

2.4 Why Is Justice Needed in Critical Mineral Development?

It seems strange to ask such a question of why justice is needed. However, it is clear that in the context of the development of critical minerals, that question needs to be asked. An example of why is the significantly documented issue of child labour surrounding cobalt extraction in the Democratic Republic of Congo.[39] One of the reasons that mining law developed was the repeated safety failures that accompanied the commercial expansion of coal mining in the early 1800s.[40] Yet, nearly 200 years later, many parts of the world continue to face similar challenges. Justice, in reality, has only touched some parts of the mining sector.

The core purpose of incorporating justice in the critical mining sector is to promote fairness, advance equality and address persistent inequalities. Economics, since the 2007–2009 financial crisis, has started to engage in this issue in a significant way.[41] Indeed, another eminent economist, Piketty (2020: 670),[42] decries the issue and states that despite living in a world of big data, public data on inequality is inadequate. Piketty[43] further highlights the relationship between inequality and climate change, noting that the economic costs could amount to 5–20% of Global GDP, particularly for developing economies, if these challenges are not addressed. Supporting this argument, he cites the Stern Review (2007) and the IPCC (2018) report, which demonstrate that the economic impacts of climate change may be exacerbated by pollution and environmental degradation.

Considering the broad role that critical minerals play in the developing the low-carbon economy, which the majority of countries have committed to underthe Paris COP21 Agreement, it is necessary to adopt a broad conception of justice that can encapsulate this global industry and its

[39] Heffron, R. J. 2018. The application of distributive justice to energy taxation utilising sovereign wealth funds. Energy Policy, 122, 649–654.

[40] Heffron, R. J. and Talus. K. 2016. The Evolution of Energy Law and Energy Jurisprudence: Insights for Energy Analysts and Researchers. Energy Research and Social Science, 19, 1–10.

[41] Scheidel, W. 2017. The Great Leveler. Princeton University Press: NJ, US; Stiglitz, J. E. 2012. The Price of Inequality. Penguin Books: London, UK; Atkinson AB. 2015. Inequality: what can be done? MA (US): Harvard University Press.

[42] P.670, Piketty, T. 2020. Capital and Ideology (translated by Goldhammer, A.). Belknap Press of Harvard University Press: MA, US.

[43] Piketty, T. 2020. Capital and Ideology (translated by Goldhammer, A.). Belknap Press of Harvard University Press: MA, US.

J	**T R A N S I T I O N**	**Justice**	**Justice takes the form of 3 forms of justice** **Distributive** **Procedural** **Restorative**
U		**Universal**	**Universal takes the form of two universal forms of justice** **Recognition** **Cosmopolitanism**
S		**Space**	**Space brings in location, where are 'events' happening ? (in principle, at local, national and international levels)**
T		**Time**	**Time brings us into transition timelines such 2030, 2050, 2080 etc. and also 'speed' of the energy transition (i.e. is it happening fast enough?).**

Fig. 2.2 The legal geography 'JUST' framework for the just transition. (Source: Adapted by Author from Geoforum)

far-reaching impacts. In this context, there is a need to incorporate elements of climate, environmental and energy justice, and these can be united together in terms of having a just transition to a low-carbon economy. This can be best expressed through the *JUST Framework* below that applies for a Just Transition[44]—see Fig. 2.2.[45] This framework unites climate, environmental and energy (CEE) justice scholarship and can increase the potential for reducing inequality and injustice within society.

The benefit of utilising the *JUST Framework* is that it enables a more practical and interdisciplinary perspective and ensures the researcher assesses the justice issues specific to the problem, wider societal justice issues in terms of universal justice, the geography issues such as where (i.e. need to consider global supply chains) and then finally to consider energy, climate and transition timelines (such as 2030, 2040, 2050, and beyond). The *JUST Framework* covers five key elements of justice that are needed to ensure the application of human rights across the energy life-cycle[46] and these are identified briefly below:

[44] Heffron R. J. and McCauley, D. 2018. What is the 'Just Transition'? Geoforum, 88, 74–77.

[45] Heffron R. J. and McCauley, D. 2018. What is the 'Just Transition'? Geoforum, 88, 74–77.

[46] Heffron, R. J. & McCauley, D. 2017. The concept of energy justice across the disciplines. Energy Policy, 105, 658–667.

- *Distributive justice*—this concerns the distribution of benefits from the energy sector and also the negatives (i.e., are oil and gas revenues shared sufficiently?; who suffers the environmental damage?);
- *Procedural justice*—the focus here is on legal process and the necessary full legal steps (i.e., are all the steps for an environmental impact statement observed?);
- *Recognition justice*—are rights recognised for different groups in society i.e., in particular are we recognising the rights of indigenous communities?;
- *Cosmopolitanism justice*—this stems from the belief we are all citizens of the world and so have we considered the effects beyond our borders and from a global context?;
- *Restorative justice*—any injustice caused by the energy sector should be rectified, and it focuses on the need for enforcement of particular laws (i.e., energy sites should be returned to former use, hence waste management policy and decommissioning should be properly done).

There is motivation to utilise the energy justice framework as this will ensure that policymaking moves a country towards a just transition to a low-carbon economy. Further, in using this framework it becomes a risk reduction policy tool. The benefits are that it applies a more systematic and holistic approach to an energy (including mining) issue. Recent literature utilising this JUST framework demonstrates its value across a range of contexts, including energy democracy,[47] gender issues,[48] critical minerals

[47] There are multiple here but we choose this one: Droubi, S., Heffron, R. J. & McCauley, D. 2022. A critical review of energy democracy: A failure to deliver justice? Energy Research & Social Science, 86, 102444.

[48] Sumarno, T. B., Sihotang, P. and Prawiraatmadja, W. 2022. Exploring Indonesia's energy policy failures through the JUST framework. Energy Policy, 164, 112914.

in the ASEAN region,[49] construction,[50] investment,[51] corporate sustainability,[52] jobs[53] and education.[54]

2.5 Applying the '*JUST*' *Framework*': Theoretical Assumptions for the Critical Minerals Industry

In utilising the *JUST* framework from Section 2.4 above, there are key theoretical assumptions which can be advanced for ensuring and measuring the level of justice necessary for critical minerals development. Each of the four core parts of the JUST framework is examined—justice, universal, space and time—and analysed in turn. The focus here concerns what the key issues are that society needs to ensure are happening as the critical mineral sector develops across the world. In some cases, there are already actions happening that the critical minerals industry can adopt. The aim is not to explore every single type of issue but instead focus on one and/or several key issues under each of the four parts of the JUST framework. Finally, the contribution here lies in how to move from theory to practice in terms of achieving justice as the critical mineral industry develops.

[49] Qurbani, I. D., Heffron, R. J. & Rifano, A. T. S. 2021. Justice and critical mineral development in Indonesia and across ASEAN. The Extractives Industry and Society, 8 (1), 355–362.

[50] Wasalkar, W., Goel, U. and Raheja, L. 2025. Just Transition in the Context of Circular Economy: Developing a Comprehensive Policy Framework with an Example of the Construction Industry. JTRC-Indian Institute of Technology Kanpur Research Paper No. 12.

[51] Heffron, R. J., Merdekawati, M., Suryadi, B. and Yurnaidi, Z. 2024. Defining a 'Just Energy Investment' for the ASEAN Just Transition. Environmental and Sustainability Indicators, 22, 100367.

[52] Hazrati, M. and Annamyer, E. 2025. Evaluating EU Directives on Corporate Sustainability Reporting and Due Diligence Through the Lens of the 'JUST' Framework. In In: Phillips, A., Kaschny, L. (eds) New Engagement Strategies for Energy Justice. Just Transitions. Palgrave Macmillan, Cham.

[53] Smart Prosperity Institute (Pai, S., Harrison, K. and Zerriffi, H. 2020. A Systematic Review of the Key Elements of a Just Transition for Fossil Fuel Workers. Clean Economy Working Paper Series, WP 20–04. Available at: https://institute.smartprosperity.ca/sites/default/files/transitionforfossilfuelworkers.pdf (last accessed 8 June 2025).

[54] Droubi, S., Galamba, A., Fernandes, F. K., de Mendonca, A. A. & Heffron, R. J. 2023. Transforming education for the just transition. Energy Research & Social Science, 100, 103090.

2.5.1 *Justice and Its Elements: Distributive, Procedural and Restorative Justice*

2.5.1.1 Distributive Justice

Distributive justice is one of the three cornerstones of justice. There is a need to ensure there is a fair distribution of resources in society. As stated earlier in Section 2.4, many of the leading economists have been working directly on this issue where there is no fair distribution of resources in society, resulting in inequalities. This issue is crucial as the critical minerals industry develops. There is a need to ensure that the same issues that have arisen in the extractives industry do not arise again.

The core issue for distributive justice is taxation. It is imperative to know where the revenue from the critical minerals is going and how it is distributed. Recent scholarship has concluded that distributive justice should be recognised as the cornerstone of tax law, i.e. as the "*first or sovereign virtue of a society's tax system*".[55] It has even been argued that Adam Smith viewed distributive justice as a key objective of tax policy, and that legal certainty was necessary to ensure that distributive justice was more than an illusory goal, a principle that remains relevant today.[56] This is undoubtedly an issue for many developing countries when establishing their taxation policy around critical minerals.

The reality, however, is that revenue distributed from an extracted resource is difficult to calculate because extraction is a complex process, coupled with the fact that prices are determined in global markets. Hence, revenue sharing arrangements for most projects are negotiated on a case-by-case basis, despite the existence of taxation legislation. There are, however, actions on this issue, through the Extractive Industries Transparency Initiative[57] (which requires a state and an investor to disclose the details of their taxation relationship). It is having an impact, though there remain significant problems, with only 53 countries having signed up and not all cooperating to the same extent.

[55] Duff, D., 2017. Tax Policy and the Virtuous Sovereign: Dworkinian Equality and Redistributive Taxation, (p.167–89). In: Bhandari, M. (Ed.), Philosophical Foundations of Tax Law 2017 Oxford University Press, Oxford, UK.

[56] Gribnau, H., Vording, H., 2017. The birth of tax law as a legal discipline. In: Harris, P., de

[57] Extractive Industries Transparency Initiative. 2020. Available at: https://eiti.org/ (last accessed, 8 June 2025)

There is further global action from an OECD-led initiative, which seeks to reform issues around taxation of energy and resource extraction.[58] In addition, this OECD initiative representing circa 60 countries noted that energy and mineral resource extraction represents a significant opportunity to raise revenue for many Governments. Post COVID-19, the need for more revenue will be even more pronounced, thus, tax reform must occur at the international and national levels. More transparency in taxation issues is needed at the international level, while at the national level, tax reforms are needed to increase revenue.

2.5.1.2 Procedural Justice

Procedural justice focuses specifically on the legal process involved in taking a project from planning and construction through to operation and end-use, and whether all stakeholders have a legitimate opportunity to be involved (or represented) at each stage. There are, of course, many issues here in terms of finance and taxation, which are mentioned in the elements of justice discussed. The key issue for the development of the critical mineral industry in procedural justice is the Environmental Impact Assessment (EIA) process that each country has at a national level.

This is the legal process a project developer has to go through in three phases and at international, national and local levels:

1. *International:* In securing finance for the project this will require an EIA to be produced for the financing institution under international banking standards—the Equator principles (this may be different for a company financing the project in-house or a national company with financing options within the country, although there would likely remain some environmental impact statement produced);

[58] (1) The International Institute for Sustainable Development (IISD) and the Organisation for Economic Co-operation and Development (OECD). 2018a. Limiting the Impact of Excessive Interest Deductions on Mining Revenue. OECD: Paris, France; (2) The International Institute for Sustainable Development (IISD) and the Organisation for Economic Co-operation and Development (OECD). 2018b. Tax Incentives in Mining: Minimising Risks to Revenue. OECD: Paris, France; and (3) The International Institute for Sustainable Development (IISD) and the Organisation for Economic Co-operation and Development (OECD). 2018c. Monitoring the Value of Mineral Exports: Policy Options for Governments. OECD: Paris, France.

2. *National:* Has to adhere to national EIA legislation and submit an Environmental Impact Statement (EIS), which has to be approved before the project receives permission to start; and
3. *Local:* The EIA process has to include several elements of public participation and involve these local stakeholders in the plans for the development of the project. There have been two recent cases in Australia and Kenya where energy projects, coal, failed to pass the EIA process. A key reason for the failure in both countries was that the EIAs lacked completeness in terms of data provision, the poor assessment of the social and environmental impacts from the existing data and the overestimation of the projects' positive economic contributions.[59]

Significantly, the legislation around EIAs is changing periodically for two main reasons: the first is that once it changes, i.e. is improved in one country, other countries follow. The reason why this happens is that the increased requirements in the EIA process reduce the risk profile of the project for all stakeholders. The second reason for change is that data capture methods are growing, so different types of environmental impact, as well as the socio-economic impact of a project, can now be measured. EIAs are becoming more and more stringent, and this is documented in the literature, it will be increasingly complex for energy projects, including critical minerals, to pass this process.

There is additional support for EIA legislation through international law under an international agreement called the Aarhus Convention of 1998—known as the Convention on Access to Information, Public Participation in Decision-Making and Access to Justice in Environmental Matters. This agreement, signed by 39 countries (but with 47 parties), allows for public participation in energy projects and access to environmental data. The result is that there is added enforcement to national EIA legislation mainly limited to the EU and neighbouring countries, since the Convention was advanced by the UN ECE. However, perhaps a more significant development has been in international finance standards, known as, the Equator Principles. These Equator Principles will play a

[59] Nogrady, B. 2019 (11 February). Landmark Australian ruling rejects coal mine over global warming, Nature, Available at: https://www.nature.com/articles/d41586-019-00545-8 (last accessed, 8 June 2025); and Herbling, D. 2019 (26 June). 'Kenya cancels environment license of $2 billion coal-power plant', Bloomberg, Available at: https://www.bloomberg.com/news/articles/2019-06-26/kenya-cancels-environment-license-of-2-billion-coal-power-plant. (last accessed, 8 June 2025).

significant role in the critical minerals industry as they cover the majority of international project finance debt within developed and emerging markets, and notably, they require the project to have passed the EIA before releasing the project finance.[60]

2.5.1.3 *Restorative Justice*

Restorative justice concerns how injustice that may be present in the energy sector should be rectified.[61] This could be in the form of revenue distribution from the project (covered principally under distributive justice). The key issue for the critical mineral industry would be that these energy sites should be returned to their former use. Hence, waste management policy and decommissioning should be completed appropriately and costed within a project, as well as the rules set out in legislation. In addition, restorative justice can aid in pinpointing where prevention needs to occur.

Of importance across the world, as critical minerals are developed, is that these sites of extraction are not abandoned. The energy sector already has a well-documented history of abandoned energy sites, which then become environmental problems.[62] In ensuring that this does not happen in the development of the critical minerals industry and that justice has a role, a 'restorative cost' needs to be built into the project's costs, and that needs to be factored in during the three phases of the relevant critical mining project as outlined below:

[60] Equator Principles. 2020. The Equator Principles. Available at: https://equator-principles.com/about/ (last accessed, 10 April 2020).

[61] (1) Heffron, R. J. & McCauley, D. 2017. The concept of energy justice across the disciplines. Energy Policy, 105, 658–667; and (2) Heffron, R. J. and Hazrati, M. 2024. The Rise of Restorative Justice in the Energy Transition and for Climate Mitigation. Springer: Heidelberg, Germany.

[62] (1) Gallucci, M. 2016 (14 January). When a Coal Company Goes Bankrupt, Who Is Left to Clean up the Mess?' International Business Times. Available at: http://www.ibtimes.com/when-coal-company-goes-bankrupt-who-left-clean-mess-2264097 (last accessed, 8 June 2025); (2) Robertson, J. 2016 (28 January). Coal Giants Abandon Unprofitable Mines, Leaving Rehabilitation under Threat. The Guardian. Available at: <https://www.theguardian.com/environment/2016/jan/29/coal-giants-abandon-unprofitable-mines-leaving-rehabilitation-under-threat (last accessed, 8 June 2025); and (3) Miller, C. G. 2005. Financial Assurance for Mine Closure and Reclamation. Available: https://www.icmm.com/document/282 (last accessed, 8 June 2025).

1. *Project Development and Construction*
The EIA process needs to include the 'restorative cost' and plans in the EIS that is prepared and which involves multiple stakeholder consultations.
2. *Project Operation*
As was previously the case, just satisfying the local community at the outset of the project is not sufficient, and the project operator needs to work with the local community for the duration of the project operation. This has been formalised into a Social-Licence-to-Operate (SLO) agreement and in some countries is recognised in legislation.[63] Again part of the cost here would be restorative as gradually the project operations reduce over time, and these should be factored into project costs.
3. Project Closure and End-of-Life Process
This is a clear restorative cost and is, in essence, the decommissioning or waste management operational cost. This is a significant area of weakness and needs to be rectified in national and international law and policy.[64]

One tool that is already in use in some countries is forms of an Energy Financial Reserve Obligation (EFRO), which can be referred to as a clean-up obligation and/or environmental bonds. In the US, under the Federal Surface Mining Control and Reclamation Act (SMCRA) 1997 (and in Australia, the Financial Assurance under the Environmental Protection Act 1994), energy companies are required to remediate the lands where mining activity has occurred. However, many companies were allowed to self-bond and therefore, when they went bankrupt there was still no finance available for meeting reclamation obligations[65] and the EFRO counters this corporate behaviour.

[63] https://www.sciencedirect.com/science/article/abs/pii/S0301420717304786

[64] Heffron, R. J. 2018c. Energy law for decommissioning in the energy sector in the 21st century. Journal of World Energy Law & Business, 11 (3), 189–195.

[65] Bloomberg, 2016 (3 December - Thomas Biesheuvel, Jesse Riseborough & Agnieszka De Sousa). Why Bankruptcy Might Be the Mining Industry's Last Best Hope. Available at: http://www.bloomberg.com/news/articles/2015-12-03/why-bankruptcy-might-be-the-mining-industry-s-last-best-hope (last accessed, 8 June 2025); and (2) ABC Australia (Sue Lannin). 2015 (18 December). China Economist Warns Major Miners May Collapse in 2016. Available at: http://www.abc.net.au/news/2015-12-17/china-economist-warns-that-iron-ore-miners-will-collapse/7037802

The EFRO has now become legislation in several countries, and it is already causing an impact on the value of an energy asset as it requires an operator to place an estimated amount of money dedicated to the decommissioning of the energy infrastructure and site into a neutral bank account (an escrow); this is to avoid the scenario where the State has to pay should the operator go bankrupt or sell on the asset to a company without the financial capacity to pay for a clean-up.

2.5.2 *Universal*

2.5.2.1 Recognition Justice

Recognition justice is about whether the rights of different groups are recognised as development happens. In particular, recognition justice in the energy sector has focused on the effect of energy developments and activities on indigenous communities.[66] There is international law that supports the rights of indigenous communities in the United Nations Declaration of the Rights of Indigenous Peoples 2007,[67] where a key part of that treaty is Article 10—*Indigenous peoples shall not be forcibly removed from their lands or territory. No relocation shall take place without the free, prior and informed consent of the indigenous people, and after agreement on just and fair compensation and, where possible, with the option of return.*[68] Indeed, it is considered international law due to the reliance on it as well as the American Declaration of the Rights and Duties of Man (1948) by international and state tribunals.[69]

These issues arise repeatedly as a result of energy projects and most recently, one legal case directly raised this issue where the foreign company abused its position, polluted sacred places and committed other serious crimes—*South American Silver Limited (Bermuda) v. the Plurinational State of Bolivia, PCA Case No. 2013–15*. The final result was in 2018,

[66] McCauley, D., Heffron, R. J. Stephan, H. and Jenkins, K. 2013. Advancing Energy Justice: The triumvirate of tenets. International Energy Law Review, 32 (3), 107–110.

[67] United Nations Declaration of the Rights of Indigenous Peoples (UNDRIP). 2007. Available at: https://www.un.org/development/desa/indigenouspeoples/declaration-on-the-rights-of-indigenous-peoples.html (last accessed, 8 June 2025).

[68] P.11, United Nations Declaration of the Rights of Indigenous Peoples (UNDRIP). 2007. Available at: https://www.un.org/development/desa/indigenouspeoples/declaration-on-the-rights-of-indigenous-peoples.html (last accessed, 8 June 2025).

[69] Phillips, J. S. 2015. The rights of indigenous peoples under international law, Global Bioethics, 26 (2), 120–127.

when the project had to be expropriated from the investor, with only sunk costs paid, which was $18.7 million instead of the $385 million claimed by the investor. The case was heard under the rules of the United Nations Commission on International Trade Law. Indeed, the UN have long since recognised this issue and had a report conducted in 2012 on *An Analysis on the Duty of the State to Protect Indigenous Peoples Affected by Transnational Corporations and Other Business Enterprises* which called for significant action by Governments to protect the rights of indigenous communities from resource extraction companies.[70]

In recognising the rights of indigenous communities, it is important that new legal structures strengthen their voice, and in that context, it is welcome that legal agreements upholding a *Social-Licence-to-Operate* (SLO) are on the rise. An SLO can be defined as a legal instrument for achieving justice and ensuring long-term sustainability in energy investments and a just transition to a low-carbon economy.[71] This issue is increasingly prominent as illustrated by the case of *Bear Creek v. Peru Bear Creek Mining Corporation v. Republic of Peru, ICSID Case No. ARB/14/2* concluded in 2017, which recognised the central challenge inherent in the absence of an SLO. Thus, it contributed to the expropriation of a silver mine in Peru due to significant mistrust between the investor and the local population. Consequently, only sunk costs of $18.2 million was paid instead of the $522.2 million claimed by the investor. It is clear from the judgement in the aforementioned case that the SLO had been broken by the investor and project operator and that led to the necessary expropriation of the asset.

In addition, there is a broadening of the concept of recognition justice, with many advocating that it should extend beyond indigenous communities to include other disadvantaged groups whose rights also require recongnition, essentially reflecting the broader issue of societal inequality. For example, Heffron et al. analyse mining projects in Colombia[72]—which were critical mineral projects—and highlight how the communities worst

[70] UN Economic and Social Council, 2012. An Analysis on the Duty of the State to Protect Indigenous Peoples Affected by Transnational Corporations and Other Business Enterprises. E/C.19/2012/3. Available at: https://www.un.org/esa/socdev/unpfii/documents/2012/session-11-e-c19-2012-3.pdf (last accessed, 8 June 2025).

[71] Heffron, R. J., Downes, L., Rodriguez, O. & McCauley, D. 2021. The emergence of the 'social licence to operate' in the extractive industries? Resources Policy, 74, 101272.

[72] Heffron, R. J., Downes, L., Rodriguez, O. & McCauley, D. 2021. The emergence of the 'social licence to operate' in the extractive industries? Resources Policy, 74, 101272.

affected are communities who are in 'extreme' poverty according to the UN standards. The question that needs to be asked is whether there is enough action to ensure these local mining communities share in the benefits of these critical minerals and whether their socio-economic rights are being respected.

The issue of human rights and business is a fast-growing area of research and practice in law,[73] and particularly so in terms of the distribution of tax revenue and its relationship to human rights. This study by the IBA followed a previous study which examined in detail the effects of tax abuse and had a major finding that tax abuses have a significant impact on the enjoyment of human rights.[74] In addition, the recent *UN Committee on Economic and Social and Cultural Rights* report highlights the role of states and businesses in determining the realisation of economic, social and cultural rights.[75]

2.5.2.2 Cosmopolitanism Justice

In the context of critical minerals development, the link to cosmopolitan justice, which is based on the core premise that we are all citizens of the world, is clear. There is a global effect to our actions, and as identified in Section 2.2 earlier, the critical minerals industry is global, and these minerals are utilised in everyday. There is a need to recognise, therefore, the multiple cross-border effects of our activities as a result of global supply and business value chains.

There is a growing recognition of this issue, and as will be outlined below, this must be realised and accounted for as the critical minerals extractive industry develops. There are several recent clear examples where there is a growing interest in taking legal action arising from cross-border

[73] International Bar Association (IBA). 2016. IBA Practical Guide on Business and Human Rights for Business Lawyers. The International Bar Association: London, UK.

[74] International Bar Association (Human Rights Institute). 2013. Tax Abuses, Poverty and Human Rights. IBA: London, UK. Available at: https://www.ibanet.org/Article/NewDetail.aspx?ArticleUid=4A0CF930-A0D1-4784-8D09-F588DCDDFEA4(last accessed, 8 June 2025). It should be noted also that this book is available to download freely.

[75] United Nations Economic and Social Council (UNESC). 2017 (11 August). Committee on Economic and Social and Cultural Rights—General comment No. 24 (2017) on State obligations under the International Covenant on Economic, Social and Cultural Rights in the context of business activities. Available at: http://docstore.ohchr.org/SelfServices/FilesHandler.ashx?enc=4slQ6QSmlBEDzFEovLCuW1a0Szab0oXTdImnsJZZVQcIMOuuG4TpS9jwIhCJcXiuZ1yrkMD%2FSj8YF%2BSXo4mYx7Y%2F3L3zvM2zSUbw6ujlnCawQrJx3hlK8Odka6DUwG3Y (last accessed, 8 June 2025).

or overseas effects. A first clear example, though not related to critical minerals, is in coal, where a Judge in Australia held that a coal mine should not receive permission to open because of the carbon dioxide that would be produced in other places in the world—*Gloucester Resources Limited v Minister for Planning [2019] NSWLEC 7.*

Indeed, disputes are on the rise in the mining sector, and increasingly, they concern some international issues.[76] A second example that partly reflects the critical mineral extractive industry is explicitly the actions by the Canadian Government, where in creating and appointing an Ombudsperson for Responsible Enterprise that will assess and investigate the actions of Canadian overseas companies focusing on human rights abuses in mining, oil and gas and garments.[77] That aside, there are three cases being taken on by foreign entities against Canadian companies in Canada for human rights abuses in their foreign sites; one in Eritrea and two in Guatemala.[78]

There are other examples of cross-border actions which can affect the critical minerals industry, and these relate to the aforementioned EIAs, Equator Principles and Tax initiatives under the first part of the JUST framework, *justice*. Additionally, the 2015 Paris Agreement has already been noted, and it is worth recalling that 189 countries have signed and ratified it (197 have signed overall).[79] That is a short space of time for so many countries to sign up but it emphasises the international cooperation in the world on achieving low-carbon economies. This international cooperation is also evident in another convention which is the Convention on Environmental Impact Assessment in a Transboundary Context, widely known as the Espoo (EIA) Convention (1991). In brief, this convention concerns the obligation on States to report to each other the effects of activities within their boundaries that may have cross border impacts.

[76] Heffron, R. J. 2018. Mining Disputes. In Scherer, M. 2018. (Editor). International Arbitration in the Energy Sector. Oxford University Press: Oxford, UK.

[77] Canadian Ombudsperson for Responsible Enterprise (CORE). 2020. Office of the Canadian Ombudsperson for Responsible Enterprise. Available at: https://core-ombuds.canada.ca/core_ombuds-ocre_ombuds/index.aspx?lang=eng (last accessed, 8 June 2025),

[78] Volterra Fietta. 2018. The UK and Canada are taking steps to regulate the area of business and Human Rights ("BHR"). Available at: https://www.volterrafietta.com/the-uk-and-canada-are-taking-steps-to-regulate-the-area-of-business-and-human-rights-bhr/ (last accessed, 8 June 2025).

[79] UNFCC. 2020. Paris Agreement—Status of Ratification. Available at: https://unfccc.int/process/the-paris-agreement/status-of-ratification (last accessed, 8 June 2025).

Although finalised in 1991, it was only in 1997, six years later, that there were enough countries to enter it into force, and there remain only 45 signatories to it.[80] The majority of signatories of the Espoo Convention are European because it was the UN Economic Commission for Europe that advanced it. Nevertheless, it shows the success of the Paris Agreement in terms of the number of countries signing up and in such a short space of time. If the Espoo Convention were more international, it could address a wider range of emerging issues, which are likely to grow over the coming decades. A final area of further research concerns access to critical minerals, and how this can be just and fair, given their location and ownership, building on early work on the development of a Fair Trade mineral supply.[81]

2.5.3 *Space*

The third component of the JUST framework is 'Space', which addresses a range of issues outlined below, drawing on key literature in geography related to space.

- The location, i.e. where the 'events' are happening
- Whether the action is happening at a local, national and international level
- The interlinks, i.e. is it part of a global supply chain, and what are the characteristics of this
- The stakeholders involved and their activities, i.e. a multinational company utilising tax services, for example.

It is imperative that the critical minerals industry is considered as a global industry. Actions that happen at a local level in terms of mining operations are all influenced by what happens at a national and international level. International taxation practice, for example, shapes how extracted minerals are taxed and in turn influences the scale, length and

[80] UN. 2020. Convention on Environmental Impact Assessment in a Transboundary Context Available at: https://treaties.un.org/Pages/ViewDetails.aspx?src=TREATY&mtdsg_no=XXVII-4&chapter=27&lang=en#1 (last accessed, 8 June 2025).

[81] Van Bockstael, S. 2018. The emergence of conflict-free, ethical, and Fair Trade mineral supply chain certification systems: A brief introduction. The Extractive Industries and Society, 5 (1) 52–55.

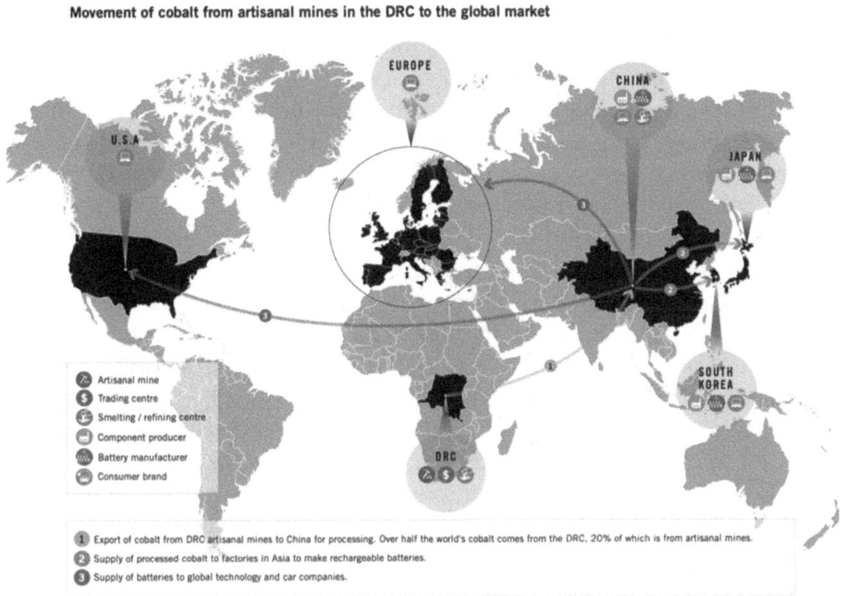

Fig. 2.3 A Global business value and supply chain for Cobalt. (Source: Amnesty International (2017))

operational design of the extraction project. In essence, actions in one part of the world influence outcomes in another.

This can be demonstrated clearly in Fig. 2.3,[82] which looks at the issue of a cobalt business value and supply chain where cobalt is used for battery technology.[83] The figure identifies three phases where the cobalt is; (1) extracted, (2) then exported to Asia, where it is utilised in battery production and (3) then eventually consumed by Western and Asian markets. In this context, the critical minerals industry has multiple locations of 'activity'.

[82]Amnesty International, 2017. Time to Recharge: Corporate Action and Inaction to Tackle Abuses in the Cobalt Supply Chain. Available at: https://www.amnesty.org/download/Documents/AFR6273952017ENGLISH.PDF (last accessed, 8 June 2025).

[83]Amnesty International, 2017. Time to Recharge: Corporate Action and Inaction to Tackle Abuses in the Cobalt Supply Chain. Available at: https://www.amnesty.org/download/Documents/AFR6273952017ENGLISH.PDF (last accessed, 8 June 2025).

As a result of this, multiple issues are raised at local, national and international levels. One of these is to explore the flow of finance around that global supply chain. The Amnesty Report (2017) highlighted the issue of labour. In essence, they state that human rights abuses are occurring in the extraction phase, so it could be determined that the DRC is not receiving its due share of revenue from cobalt. Heffron identifies the apparent injustice in terms of the tax rate, which was 6% (between 2011 and 2014), when the recommendation by the World Bank for such minerals was 46%.[84]

Further, it is important to consider where the profits from the mining sector and, as a result, where profits from the critical minerals sector will end up. This will generally be in bank accounts offshore that belong to a collection of technology companies and mining companies via the services of professional firms that assist in this process. Indeed, the IGF-OECD initiative, representing circa 60 countries, released three reports on the aggressive tax avoidance by multinational firms in the mining sector.[85] There is no doubt that the international community is thinking about how the critical minerals industry may evolve here and that confirms why it is so necessary to ensure justice within the development of the critical minerals industry.

2.5.4 *Time*

Time frames the transition through timelines such as 2030, 2040, 2050, 2080 etc., and also raises questions as to the 'speed' of the energy transition, i.e. *is it happening fast enough?*; which many argue it is not,[86] and highlights the need to focus on critical minerals, which has a crucial

[84] Heffron, R. J. 2018. The application of distributive justice to energy taxation utilising sovereign wealth funds. Energy Policy, 122, 649–654.

[85] (1) The International Institute for Sustainable Development (IISD) and the Organisation for Economic Co-operation and Development (OECD). 2018a. Limiting the Impact of Excessive Interest Deductions on Mining Revenue. OECD: Paris, France; (2) The International Institute for Sustainable Development (IISD) and the Organisation for Economic Co-operation and Development (OECD). 2018b. Tax Incentives in Mining: Minimising Risks to Revenue. OECD: Paris, France; and (3) The International Institute for Sustainable Development (IISD) and the Organisation for Economic Co-operation and Development (OECD). 2018c. Monitoring the Value of Mineral Exports: Policy Options for Governments. OECD: Paris, France.

[86] Figueres, C. et al. 2017. Three Years to Safeguard our Climate. Nature. 546 (7660) 593–595. (29 June 2017).

role to play.[87] Transition timelines and/or energy and climate timelines have become a feature of the energy system for the development of a low-carbon economy. Therefore, they have an immediate and significant effect on the critical minerals industry as they indicate which minerals will be in demand for the transition. A key influence on future timelines has been the success of the Paris COP21 climate change negotiations resulting in the Paris Agreement (2015).

The Paris Agreement, while it does not specifically bind a country to meeting a goal in 2030, it requires them to produce a clear pathway of how they will achieve these goals, giving birth to nationally determined contributions (NDCs). Some countries have been taking progressive action in meeting 2030 goals, such as the UK phasing out the use of coal by 2025[88]—but many others have 2030 goals that align with 2040 and 2050 goals, and the Paris Agreement is the first stage of this transition to 2050 (with an estimated 77 countries aiming for net zero-carbon emissions by 2050).[89]

In thinking of how to achieve, for example, the 2030 climate and energy goals, a country must establish new law on how to achieve this energy transition. There is no standard solution as countries have sovereign control over their own energy resources (a right of a nation in international law),[90] together with different geographies, cultures and socio-economic characteristics. However, there is a common demand for the critical minerals needed to transition to a low-carbon economy.

Timelines are of vital importance to the transition to a low-carbon economy and, from a legal perspective, requisite law must be enacted to deliver on climate and energy goals for 2030, 2040 and 2050. In different countries, the formulation of law can take different lengths of time for example, from one to five years, and that is not even to question whether

[87] Ali Saleem H., Damien Giurco, Nicholas Arndt, Edmund Nickless, Graham Brown, Alecos Demetriades, Ray Durrheim, and others. 2017. "Mineral Supply for Sustainable Development Requires Resource Governance." Nature 543 (March 16): 367–72.

[88] UK Government. 2016. Coal generation in Great Britain: The pathway to a low-carbon future. Available at: https://www.gov.uk/government/consultations/coal-generation-in-great-britain-the-pathway-to-a-low-carbon-future (last accessed, 8 June 2025).

[89] International Institute for Sustainable Development. (IISD) (Kosolapova, E.). 2019 (24 September). 77 Countries, 100+ Cities Commit to Net Zero Carbon Emissions by 2050 at Climate Summit. Available at: http://sdg.iisd.org/news/77-countries-100-cities-commit-to-net-zero-carbon-emissions-by-2050-at-climate-summit/ (last accessed, 8 June 2025).

[90] Heffron, R. J. 2018b. Mining Disputes. In Scherer, M. 2018. (Editor). International Arbitration in the Energy Sector. Oxford University Press: Oxford, UK.

that law is effective. The essence of the issue here is that to achieve a 2030 energy policy goal that will deliver a transition to a low-carbon economy, the law needs to be changed, reformed, etc., at least a reasonable length of time before that—one would need to think before 2022 at the latest to achieve 2030 goals. A key reason why this is necessary is to ensure that there is a stable legal structure around the activity, in this case, critical mineral development. This legal stability (or what can be referred to as *legal certainty*) provides confidence to investors who can then make an investment decision knowing that they will be able to secure a return on their investment and that their investment is protected (to some degree).

Therefore, for critical minerals to be developed in a just way, there is a need for legislative development that; (1) covers all issues around extraction, production and export and (2) ensures that the latter legislative development complies with and adapts to (such as adapting to the demand for critical minerals as a result of these international agreements) goals of the Paris Agreement at national levels and international energy and climate targets.

2.6 Conclusion: Resolving Justice for the Critical Minerals Industry

The critical minerals industry is growing at a fast pace. This chapter examined the underexplored role of justice in the development of this industry. As societies move towards being low-carbon, it is necessary that the societal problem of inequality is addressed and that there is a 'just' transition to a low-carbon economy. This research addresses these issues and, from a conceptual basis, using key elements of justice—distributive, procedural, restorative, recognition and cosmopolitan—utilising a JUST framework approach from the interdisciplinary area of legal geography. Future research can utilise this framework in forthcoming projects such as the planned uranium mine project (Greenland Mineral Ltd.'s Kuannersuit/Kvanefjeld Rare Earth-Uranium Project, Greenland).

This interdisciplinary approach aims to facilitate a more practical focus from the conceptual analysis. Hence, the paper takes a practical approach to examining the key issues under each element of justice that need to be addressed within the critical minerals industry. To ensure a role for justice in the development of the critical minerals industry, there is a need at the international, national and local levels, to address problems on taxation, environmental impact assessments, waste management, social licence to operate and cross-border actions and cooperation. These issues are time-sensitive, and if energy and climate goals for a low-carbon economy are to be achieved by 2030, 2040 or 2050, legal and policy action must begin now.

Given the current state of the world due to COVID-19 and the resulting economic crisis, there is a need for society to ensure a more equal distribution of resources. This should, in particular, happen in terms of critical minerals, and the revenue should be more equally shared across the business value and supply chains for these resources. Technology companies from the developed world should not be making what economists refer to as super-normal profits when vital components for their technology are being sourced with minimal cost and limited transaction costs, i.e. taxes, for acquiring the resource. Indeed, as recent scholarship by two leading international economists on taxation argues, research should identify a multiplicity of options for a fairer future, and they advocate for *tax justice* as a means of reducing inequality.[91] Similarly, this research has aimed to provide a multiplicity of options on how justice can be increased in the development of the critical minerals industry. A final conclusion is that, at the heart of these options, is the need (as these aforementioned economists have noted) to ensure that tax justice is applied to the critical minerals industry and that there is a fair distribution of the benefits derived from these resources.

[91] Saez, E. and Zucman, G. 2019. The Triumph of Injustice: How the Rich Dodge Taxes and How to Make them Pay. W. W. Norton & Co.: New York, US.

CHAPTER 3

Economics of the Global Mining Industry

Abstract Responsible mining can drive economic growth, reduce poverty and improve social well-being in resource-rich countries. Minerals contribute significantly to exports, government revenue and foreign direct investment, as seen in Botswana's diamond industry and copper-dependent economies like Zambia and Chile. The global transition to a low-carbon economy has sharply increased demand for critical minerals such as lithium, cobalt, copper and nickel. Demand for energy transition minerals is projected to nearly quadruple by 2030 due to renewable energy and electric vehicle growth. Mining economics is a multidisciplinary field covering microeconomic project evaluation and macroeconomic national impacts. Poor management of mining wealth can lead to the 'resource curse', marked by corruption, volatility and inequality. Commodity markets are cyclical, with boom-bust price patterns driven by global demand, such as China's industrialisation and COVID-19 recovery. Recent structural changes, including clean energy goals, supply chain shifts and resource nationalism, are reshaping the industry. Environmental, Social and Governance (ESG) considerations are now central to mining investment, valuation and corporate strategy. The power of mining economics can

This chapter was written with contribution from Godfred Kwaku Ennin.

© The Author(s), under exclusive license to Springer Nature Switzerland AG 2026
R. J. Heffron et al., *Revolution in the Global Critical Minerals Sector*, Just Transitions,
https://doi.org/10.1007/978-3-032-26366-7_3

result in it equipping all stakeholders to make informed, sustainable decisions amid evolving political, economic and environmental pressures.

Keywords Mining economics • Critical minerals • Fiscal revenue • Value capture • Commodity price volatility • Global supply chains • Resource nationalism • Beneficiation strategies • Investment frameworks • Taxation in mining • Governance and institutional capacity • African mining economies • Economic development • Just transition

3.1 Introduction

When done responsibly, mining can transform communities, improve social well-being and reduce poverty. Minerals account for a significant portion of exports, government revenue and foreign direct investment (FDI) for most countries. For example, diamond mining, which once accounted for more than 25% of Botswana's GDP, revolutionised the country's economy and now contributes to about 85% of export income.[1] Similarly, copper mining is synonymous with Zambia, Chile and the Democratic Republic of Congo (DRC), among others. Global climate action has further increased the strategic value of mineral resources. The transition to a low-carbon economy relies heavily on critical and rare minerals, such as lithium, cobalt, nickel, copper and graphite. These resources are crucial for renewable energy technologies, including wind turbines, solar panels and energy storage systems. Projections indicate that critical energy transition minerals, such as lithium, cobalt and copper, could increase almost fourfold by 2030.[2]

Mining economics, a multidisciplinary field that includes geology, finance, macroeconomic policy and sustainability, is at the heart of this transition. Mining spans across both microeconomic and macroeconomic

[1] World Diamond Council, *Diamond Empowerment, the Botswana Case Study: A Mix of Serendipity and Prudent Management* (2023) https://www.worlddiamondcouncil.org/diamond-empowerment-the-botswana-case-study/#:~:text=Today%2C%20while%20the%20diamond%20industry,was%20never%20a%20foregone%20conclusion

[2] United Nations Conference on Trade and Development (UNCTAD), *The State of Commodity Dependence 2023* (UNCTAD/DITC/COM/2023/3, United Nations 2023) https://unctad.org/system/files/official-document/ditccom2023d3_en.pdf; International Energy Agency (IEA), *Mineral Requirements for Clean Energy Transitions* (2024) https://www.iea.org/reports/the-role-of-critical-minerals-in-clean-energy-transitions/mineral-requirements-for-clean-energy-transitions accessed 23 June 2025.

branches of economics.[3] At the micro level, it involves cost modelling, revenue forecasting, capital budgeting and financial and risk assessment: evaluating mine projects using financial metrics such as Net Present Value (NPV), Internal Rate of Return (IRR), and All-In Sustaining Costs (AISC). At the macro level, it considers the impact of mining on fiscal revenues, trade balances, exchange rates and national employment, among others.[4] Countries that struggle to manage these dynamics effectively often fall into the proverbial 'resource curse syndrome', where resource wealth fuels corruption, economic volatility and social inequality, stifling long-term growth.

The economics of mining are primarily influenced by long periods of commodity price volatility, characterised by alternating boom-bust cycles (Fig. 3.1). For example, the commodity super-cycle in the early 2000s[5] was fuelled by rapid industrialisation in China,[6] which drove up prices for iron ore, copper and coal. Around this time, copper prices soared, prompting a surge in exploration and mine development.[7] However, according to Stocker et al.,[8] a steep decline from 2014 to 2016 followed this upswing, highlighting the weaknesses of operations that were overly leveraged and lacked sufficient capital discipline. Companies then streamlined and

[3] Ehimika Oshionebo, *Mineral Mining in Africa: Legal and Fiscal Regimes* (Routledge 2020); RF Mikesell and JW Whitney, *The World Mining Industry: Investment Strategy and Public Policy* (Routledge 2017).

[4] Natural Resource Governance Institute, *Fiscal Regime Design: What Revenues the Government Will Be Entitled to Collect* (2015) https://resourcegovernance.org/sites/default/files/nrgi_Fiscal-Regime-Design.pdf accessed 3 July 2025

[5] Martin Stocker, John Baffes and Dana Vorisek, 'What Triggered the Oil Price Plunge of 2014–2016 and Why It Failed to Deliver an Economic Impetus in Eight Charts' (World Bank Blogs, 2018) https://blogs.worldbank.org/en/developmenttalk/what-triggered-oil-price-plunge-2014-2016-and-why-it-failed-deliver-economic-impetus-eight-charts accessed 23 July 2025.

[6] FP Erdem and İ Ünalmış, 'Revisiting Super-Cycles in Commodity Prices' (2016) 16(4) *Central Bank Review* 137.

[7] O Canuto, 'The Commodity Super Cycle: Is This Time Different?' (2014) 150 *Economic Premise* 1.

[8] M Stocker, J Baffes and D Vorisek, 'What Triggered the Oil Price Plunge of 2014–2016 and Why It Failed to Deliver an Economic Impetus in Eight Charts' (World Bank Blogs, 2018)https://blogs.worldbank.org/en/developmenttalk/what-triggered-oil-price-plunge-2014-2016-and-why-it-failed-deliver-economic-impetus-eight-charts

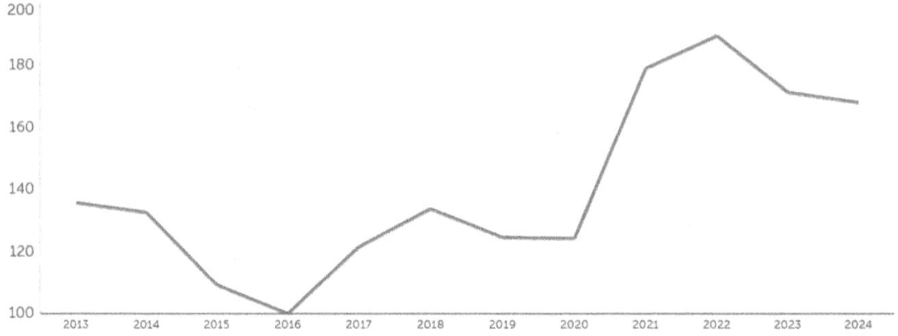

Fig. 3.1 Energy transition metal index. (Source: IMF (2025))

consolidated their cost-bases.[9] The industry was put to the test once more during the COVID-19 pandemic. Maoela et al. demonstrate that, despite brief slowdowns caused by supply chain interruptions and lockdowns, the industry showed resilience.[10] Due to the stimulus package introduced by various governments and a resurgence in demand for industrial raw materials, prices for gold, iron ore and many battery metals experienced a significant recovery by the end of 2020 and the beginning of 2021.[11] Prices for three major battery metals (nickel, lithium and cobalt), in particular, skyrocketed as demand from EV manufacturers increased.[12]

[9] H Gupta, S Periwal, O Ramsbottom and J Whitecross, 'How to Navigate Mining's Cash-Flow Conundrum' (McKinsey & Company, 22 February 2022) https://www.mckinsey.com/industries/metals-and-mining/our-insights/how-to-navigate-minings-cash-flow-conundrum accessed 23 July 2025.

[10] MA Maoela, L Chapungu and G Nhamo, 'The Socio-Economic Impacts of COVID-19 on Selected Mining Companies in Limpopo Province, South Africa' (2024) 18 *The Extractive Industries and Society* 101462.

[11] N Conte, 'How Metals Prices Performed in 2021' (Visual Capitalist, 2022) https://elements.visualcapitalist.com/how-metals-prices-performed-in-2021/ accessed 3 July 2025; M Parizot, 'Commodities Jump in 2021 as Economies Re-Open' (CIM Magazine, 2021) https://magazine.cim.org/en/news/2021/commodities-jump-in-2021-as-economies-re-open-en/ accessed 23 July 2025.

[12] M Ferguson, 'COVID-19 Impact & Recovery: Metals and Mining Outlook for H2 2021' (S&P Global Market Intelligence, 2021) https://www.spglobal.com/market-intelligence/en/news-insights/research/covid-19-impact-recovery-metals-mining-outlook-for-h2-2021 accessed 3 July 2025.

Presently, new structural forces are redefining mining economics. The clean energy transition,[13] the reconfiguration of global supply chains,[14] coupled with increasing resource nationalism,[15] are all raising the strategic importance of mineral extraction. According to the IEA (2024), without significant investment in new mining capacity, the global supply of critical minerals is expected to fall short of projected demand by 2030, potentially delaying climate objectives. At the same time, the sector faces growing scrutiny over its environmental and social practices. Environmental, Social and Governance (ESG) compliance, once peripheral, is now central to investment decisions, project licensing and permitting, and overall corporate strategy.[16] Consequently, mining economics now requires a broader and more dynamic framework—one that incorporates ESG considerations into valuation and planning.

This divergence is evident in the rise of sustainability-adjusted project finance, real options valuation techniques and ESG-induced debt instruments. Mining companies are increasingly incorporating carbon pricing into their internal models and aligning with global sustainability frameworks, such as the Task Force on Climate-related Financial Disclosures (TCFD) and the Sustainability Accounting Standards Board (SASB). Leading multinational firms like BHP, Rio Tinto and Anglo American are embedding ESG metrics into board-level decision-making and executive compensation.[17] On the other hand, governments are also re-examining

[13] J Zadeh, 'Energy Transition's Profound Impact on Mining Economics' (Discovery Alert, 2025) https://discoveryalert.com.au/news/energy-transition-mining-renewable-clean-2025/ accessed 3 July 2025.

[14] S Emran and O Tayebi, '"When Elephants Fight", Africa Transforms: The New Geopolitics of Critical Mineral Supply Chains and the Emergence of Local Value Chains' (Policy Center for the New South, Policy Paper No PP-01/24, 2024); S Dou, Y Zhu, J Liu and D Xu, 'The Power of Mineral: Shock of the Global Supply Chain from Resource Nationalism' (2024) 184 World Development 106758.

[15] R Mills, 'A Vicious Cycle of Rising Resource Nationalism' (Mining.com, 2023) https://www.mining.com/web/a-vicious-cycle-of-rising-resource-nationalism/ accessed 23 July 2025.

[16] Maybee, B., Lilford, E., & Hitch, M. (2023). Environmental, Social and Governance (ESG) risk, uncertainty, and the mining life cycle. *The Extractive Industries and Society, 14*, 101244; Bandura, R., & Hardman, A. (2023). *Environmental, social, and governance best practices applied to mining operations.* Center for Strategic and International Studies. https://www.csis.org/analysis/environmental-social-and-governance-best-practices-applied-mining-operations

[17] Kristoff, M. (2019). *BHP links executive pay to Scope 3 emission reductions, addresses thermal coal.* Institute for Energy Economics and Financial Analysis. https://coilink.

their fiscal policies and offering incentives, such as tax exemptions or expedited permitting for essential minerals. Some have implemented Production-Sharing Agreements (PSAs) and progressive royalty schemes,[18] while others have introduced export restrictions and other retroactive taxes, which can erode investor confidence.[19]

Mining economics provides stakeholders with the analytical tools they need to make responsible, strategic and well-informed decisions in the rapidly evolving modern world. It gives stakeholders a glimpse into a complex world where politics and policy intersect, sustainability and regulation sometimes conflict and finance and geology converge. The next sections explore these themes in greater depth, beginning with the core economic principles that shape mineral extraction and advancing through market dynamics, finance, taxation, technological innovation and strategic adaptive systems.

3.2 Mineral Extraction Economic Principles

3.2.1 Scarcity, Depletion and Resource Economics

The finite nature of mineral deposits elevates their value and drives the need for careful economic planning. Scarcity transcends the physical availability of these mineral deposits: it reflects economic accessibility, driven primarily by technological and policy considerations.[20] Resources that exist geologically but cannot be mined profitably due to depth, low grade or remoteness are not economically scarce; instead, they are technically inaccessible under current market conditions.

org/20.500.12592/8x2v70; Rathobei, K. E., Ranängen, H., & Lindman, Å. (2024). Exploring broad value creation in mining-Corporate social responsibility and stakeholder management in practice. *The extractive industries and society*, *17*, 101412.

[18] Juranek, S., Schindler, D., & Schneider, A. (2023). Royalty taxation under tax competition and profit shifting. *Canadian Journal of Economics/Revue canadienne d'économique*, *56*(4), 1377–1412; Castillo, E. (2021). The impacts of profit-based royalties on early-stage mineral exploration. *Resources Policy*, *73*, 102231.

[19] Acheampong, T. (2024) *From mines to markets: How Africa and Europe can become green industry partners of choice*, *ECFR*. Available at: https://ecfr.eu/publication/from-mines-to-markets-how-africa-and-europe-can-become-green-industry-partners-of-choice (Accessed: 23 June 2025).

[20] Lööw, J., Abrahamsson, L., & Johansson, J. (2019). Mining 4.0—The impact of new technology from a work place perspective. *Mining, Metallurgy & Exploration*, *36*, 701–707.

The Hotelling Rule remains a cornerstone of natural resource economics[21] and utilisation.[22] In essence, the value of a finite resource should increase over time, in line with the rate of interest, assuming no changes in technology or discovery.[23] Although real-world mineral markets deviate from this ideal path due to factors such as price fluctuations, new discoveries and unpredictable demand, the rule suggests that resource depletion will ultimately drive producers to explore more costly sources or alternatives over time.[24] This growing resource scarcity, as predicted by the Hotelling rule, is reflected in the long-term decline in ore grades.[25] The economic consequence is that more rock must be moved, processed and managed to extract the same amount of metal, thereby raising economic costs,[26] as well as time, energy and environmental challenges in most instances.[27]

3.2.2 Cost Structures: Fixed, Variable and Marginal Costs

Mining operations typically involve a combination of both fixed and variable costs. Fixed Costs or Capital Expenditures (CAPEX), are incurred upfront and include exploration, feasibility studies, infrastructure and equipment. These costs are largely sunk, meaning they cannot be recovered if a project is abandoned. Variable Costs or Operational Expenditures

[21] Gordon, R. L., & Tilton, J. E. (2008). Mineral economics: Overview of a discipline. *Resources policy*, *33*(1), 4–11.

[22] Franco, M. P., Gaspard, M., & Mueller, T. (2019). Time discounting in Harold Hotelling's approach to natural resource economics: The unsolved ethical question. *Ecological Economics*, *163*, 52–60.

[23] Livernois, J. (2009). On the empirical significance of the Hotelling rule. *Review of Environmental Economics and policy*. 3(1), https://doi.org/10.1093/reep/ren017; Mlambo, C. (2022). Non-renewable resources and sustainable resource extraction: an empirical test of the hotelling rule's significance to gold extraction in South Africa. *Sustainability*, *14*(17), 10619.

[24] Mlambo, C. (2022). Non-renewable resources and sustainable resource extraction: an empirical test of the hotelling rule's significance to gold extraction in South Africa. *Sustainability*, *14*(17), 10619.

[25] Calvo, G., Mudd, G., Valero, A., & Valero, A. (2016). Decreasing ore grades in global metallic mining: a theoretical issue or a global reality?. *Resources*, *5*(4), 36.

[26] Baker, C. (2024). *The rising cost of energy for mining*. African Mining Online. https://www.africanmining.co.za/2024/02/20/the-rising-cost-of-energy-for-mining/

[27] Calvo, G., Palacios, J. L., & Valero, A. (2022). The influence of ore grade decline on energy consumption and GhG emissions: The case of gold. *Environmental Development*, *41*, 100683.

(OPEX), however, fluctuate with production levels and include energy, labour, maintenance and consumables such as explosives and reagents.

Many mining projects are highly sensitive to commodity price fluctuations, thus, maintaining cost discipline is essential for long-term sustainability and business survival.[28] As mineral deposits become harder to reach or contain lower ore grades,[29] mining operations will continue to face rising marginal costs (i.e., the cost of producing one additional unit of output).[30] Marginal cost, in essence, plays a critical role in determining pricing thresholds, production decisions and supply responses.[31] If market prices fall below the marginal cost, the operation becomes unprofitable, and companies may halt production until favourable conditions arise. For example, several mines were put into care and maintenance during the COVID-19 pandemic as the downturn in the global demand increased production costs.[32]

Among the most significant components of mining costs is energy.[33] Depending on the type of operation and location, energy and fuel costs account for approximately 50% of a mine's total expenses.[34] According to Maennling and Toledano, energy expenses in mining account for about 15–40% of the OPEX.[35] Underground mines, for example, consume large quantities of electricity for ventilation and pumping, while open-pit mines

[28] PWC. (2018). *Mine 2018: Tempting times.* PricewaterhouseCoopers. https://www.pwc.com/id/en/publications/assets/eumpublications/mining/mine-2018.pdf

[29] Ghorbani, Y., Nwaila, G. T., Zhang, S. E., Bourdeau, J. E., Cánovas, M., Arzua, J., & Nikadat, N. (2023). Moving towards deep underground mineral resources: Drivers, challenges and potential solutions. *Resources Policy*, 80, 103222.

[30] Carlin, D., & Arshad, M. (2024). *Climate Risks in the Metals and Mining Sector* (Sectoral Risk Briefings: Insights for Financial Institutions). UNEP Finance Initiative. https://www.unepfi.org/wordpress/wp-content/uploads/2024/05/Climate-Risks-in-the-Metals-and-Mining-Sector-1.pdf

[31] Schramm, G. (1991). Marginal cost pricing revisited. *Energy economics*, *13*(4), 245–249.

[32] Maphatsoe, K. (2020) *Pandemic pushes more mines into care and maintenance*, *Mining Weekly*. Available at: https://www.miningweekly.com/article/pandemic-pushes-more-mines-into-care-and-maintenance-2020-10-28 (Accessed: 23 June 2025).

[33] Igogo, T., Lowder, T., Engel-Cox, J., Newman, A. M., & Awuah-Offei, K. (2020). Integrating clean energy in mining operations: opportunities, challenges, and enabling approaches.

[34] Patyk, M., & Bodziony, P. (2024). Empirical Analysis of Mining Costs Amid Energy Price Volatility for Secondary Deposits in Quarrying. *Energies*, *17*(3), 718.

[35] Maennling, N., & Toledano, P. (2018). The renewable power of the mine. *Available at SSRN 3661616.*

depend on fuel-intensive haul trucks and crushers.[36] Additionally, in remote regions, where grid access is unavailable, companies must rely on diesel generators or invest in renewable microgrids to reduce fuel costs and emissions.[37]

This cost exposure has made All-In Sustaining Costs (AISC) a preferred metric among analysts and investors, particularly in the gold sector.[38] Developed by the World Gold Council in 2013, AISC encompasses not only cash costs but also sustaining capital, closure obligations and general administrative expenses, which provides a more comprehensive view of a project's economic durability and lifespan (Fig. 3.2).

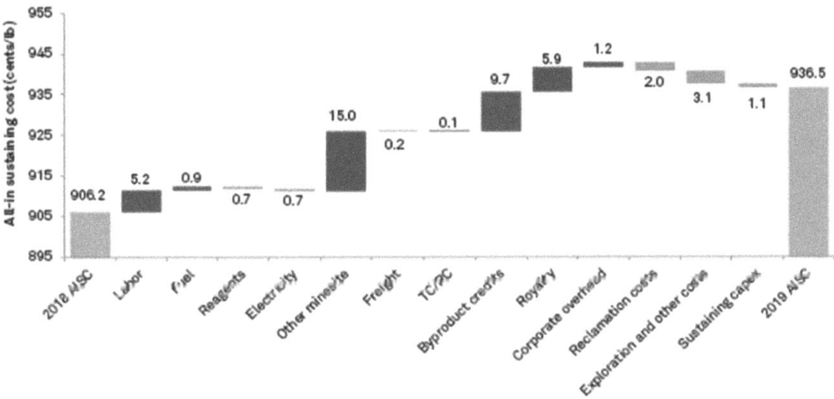

Fig. 3.2 Example of Gold All-in Sustaining Cost (AISC). (Source: Lanuza, 2020)

[36] Yeboah, D., Ackor, N., & Abrowah, E. (2023). Evaluation of wind energy recovery from an underground mine exhaust ventilation system. *Journal of Engineering*, *2023*(1), 8822475.

[37] Enemuo, M., & Ogunmodimu, O. (2025). Transitioning the mining sector: A review of renewable energy integration and carbon footprint reduction strategies. *Applied Energy*, *384*, 125484.

[38] Yapo, A. G., & Camm, T. W. (2017). *All-in sustaining cost analysis: Pros and cons.* Mining Engineering Faculty, Montana Tech of the University of Montana. Digital Commons @ Montana Tech.

3.3 Demand Elasticity and Supply Responses

Demand dynamics also play a central role in the economics of mineral extraction, as mineral resource prices become increasingly susceptible to demand levels.[39] The price elasticity of demand (the degree of responsiveness of consumers to changes in price) tends to be low in the short term for most industrial metals.[40] This is because end-user sectors such as construction, automotive and electronics cannot easily substitute materials or change production methods. Over the longer term, however, sustained high prices can incentivise substitution or recycling. The shift from platinum to palladium and rhodium in vehicle catalytic converters during the 2010s is one such example.[41]

Supply responses, in contrast, are typically inelastic in the short term. According to Diene et al., developing a mine from discovery to production can take an average of 17 years.[42] This lengthy lead time means that sudden increases in demand often outpace supply growth, leading to automatic price hikes. The lithium market provides a recent case in point. IEA reports that demand for EV batteries reached more than 750 GWh in 2023, up 40% relative to 2022, though the annual growth rate slowed slightly compared to 2021–2022.[43] Similarly, according to S&P Global, critical energy transition technologies will require more copper than their

[39] Yu, D., Wenhui, X., Anser, M. K., Nassani, A. A., Imran, M., Zaman, K., & Haffar, M. (2023). Navigating the global mineral market: A study of resource wealth and the energy transition. *Resources Policy*, *82*, 103500.

[40] Farooki, M. Z. (2010). *The Southern engine of growth and hard commodity prices: Does China lead to disruptive development.* Open University (United Kingdom); Sanyal, R. K., Mukhopadhyay, N., & Mitra, A. (2021). Demand and supply cross-explanation and their magnitude in changing open market economy. *International Journal of Recent Scientific Research*, *12*(3), 41163–41171.

[41] Rinkovec, J. (2019). Platinum, palladium, and rhodium in airborne particulate matter. *Arhiv za higijenu rada i toksikologiju*, *70*(4), 224–230; World Platinum Investment Council. (2022). *Substitution among platinum group metals.* CME Group. https://www.cmegroup.com/education/articles-and-reports/substitution-among-platinum-group-metals.htm

[42] Diene, P. D., Manley, D. A. V. I. D., Olan'g, S. I. L. A. S., & Scurfield, T. H. O. M. A. S. (2022). *Triple Win: How Mining Can Benefit Africa's Citizens, Their Environment and the Energy Transition.* Natural Resource Governance Institute.

[43] IEA (2024b), Global Critical Minerals Outlook 2024, IEA, Paris https://www.iea.org/reports/global-critical-minerals-outlook-2024, Licence: CC BY 4.0IEA (2024), Global EV Outlook 2024, IEA, Paris https://www.iea.org/reports/global-ev-outlook-2024, Licence: CC BY 4.0

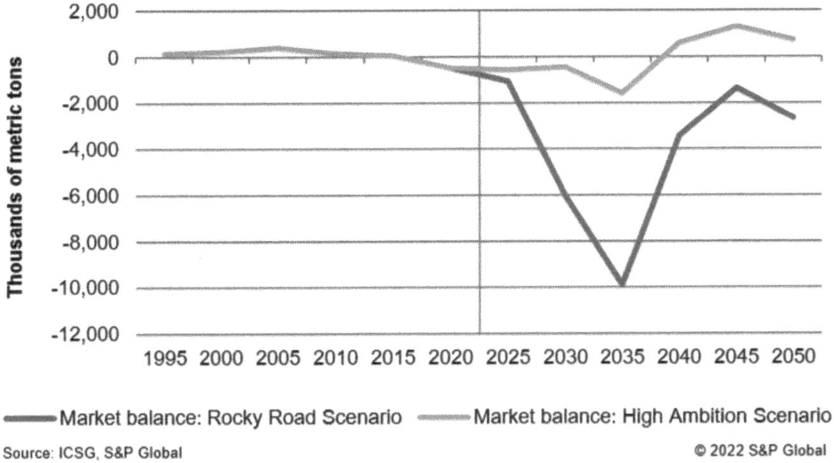

Fig. 3.3 Global copper market balance. (Source: S&P Global Market Intelligence, 2022: p12)

fossil-based counterparts.[44] As a result, the demand for copper is projected to double by 2050. Nolan reports that copper demand is expected to double by 2035 and surge even higher by 2050.[45] The rise in demand without a corresponding increase in supply naturally forces prices up (Fig. 3.3).[46]

In addition to long lead times, geopolitical factors further complicate supply elasticity.[47] Concentration of production in a few countries can expose markets to disruptions from political instability, export restrictions or trade disputes. Cobalt production, for instance, is heavily concentrated in the DRC, whereas China dominates rare earth processing and graphite

[44] Lanuza, E. (2020). *S&P Global Market Intelligence, 2021 Projection: Gold All-in Sustaining Costs | S&P Global Market Intelligence.* Available at: https://pages.marketintelligence.spglobal.com/gold-in-all-sustaining-costs-projection-2021.html (Accessed: 23 June 2025).

[45] Nolan, R. (2024). *Copper: Powering America's future and securing our supply chains.* National Mining Association. https://nma.org/2024/10/17/copper-powering-america/

[46] Sanyal, R. K., Mukhopadhyay, N., & Mitra, A. (2021). Demand and supply cross-explanation and their magnitude in changing open market economy. *International Journal of Recent Scientific Research, 12*(3), 41163–41171.

[47] Qin, Y., & Zhang, Z. (2024). Transmission mechanisms of the effects of geopolitical risk on energy returns and volatility. *International Review of Financial Analysis, 95*, 103363.

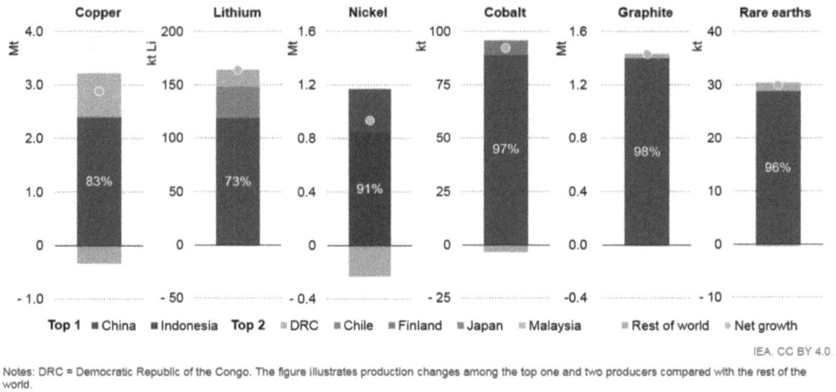

Fig. 3.4 Change in refined material production by country, 2020–2024. (Source: (UNCTD, 2023))

refining—Fig. 3.4. These geopolitical bottlenecks have prompted countries such as the United States, Canada, the European Union and Australia to pursue supply chain diversification, including in Africa.[48]

3.4 Mining Market Structures and Price Formation

In simpler markets such as those for construction materials like sand, gravel and crushed stone, competition tends to be intense.[49] The reason is that these commodities are low-value relative to their weight and are typically consumed near their source due to high transportation costs.[50] Consequently, these markets are predominantly localised, with many small-scale producers competing primarily on price and delivery logistics. The products are homogenous, and barriers to entry are relatively low,

[48] Acheampong, T. (2024) *From mines to markets: How Africa and Europe can become green industry partners of choice, ECFR*. Available at: https://ecfr.eu/publication/from-mines-to-markets-how-africa-and-europe-can-become-green-industry-partners-of-choice (Accessed: 23 June 2025).

[49] Bush, A. L. (2001). Construction materials: lightweight aggregates. *Encyclopedia of materials: science and technology*, 1550–1558.

[50] Danso, H., & Obeng-Ahenkora, N. K. (2018). Major determinants of prices increase of building materials on Ghanaian construction market. *Open Journal of Civil Engineering*, *8*(2), 142–154.

resulting in some perfect competition.[51] Here, firms act as price takers, with little ability to influence the final market prices, and success often hinges on cost efficiency and location.

As one moves up the value chain to industrial minerals like gypsum, salt and clay, markets remain relatively competitive but begin to show signs of monopolistic competition.[52] Producers may differentiate based on quality, processing methods or environmental performance, gaining a modest degree of pricing power. It is worth noting that in these markets, brand reputation, consistency of supply and customer service become more important for sustaining market share and value.[53]

In contrast, many base (iron ore, copper, aluminium, zinc, etc.), battery (cobalt, nickel, lithium, graphite, rare earths, etc.), and precious metal (silver, platinum, etc.) markets operate under oligopolistic conditions,[54] where a smaller number of large firms control a significant portion of global supply. These firms often own vast, vertically integrated operations that span exploration, mining, processing, transport and even downstream product manufacturing.[55] Participating in these markets requires substantial capital, as well as technical and regulatory expertise to operate efficiently and in compliance with environmental standards.[56] For example, Löf and Löf show that in 2020 about 60% of seaborne iron ore trade was controlled by just three companies, Rio Tinto, BHP and Vale, each with

[51] Gjerstad, S. (2007). The competitive market paradox. *Journal of Economic Dynamics and Control*, *31*(5), 1753–1780; Triani, N. V. (2023). *Perfect competition market*. SSRN. https://doi.org/10.2139/ssrn.4338530

[52] Todorova, T. (2021). Some Efficiency Aspects of Monopolistic Competition: Innovation, Variety and Transaction Costs. *Theoretical and Practical Research in Economic Fields (TPREF)*, *12*(24), 82–88.

[53] Gonen, L. D., Tavor, T., & Spiegel, U. (2024). Unlocking Market Potential: Strategic Consumer Segmentation and Dynamic Pricing for Balancing Loyalty and Deal Seeking. *Mathematics*, *12*(21), 3364.

[54] Valaskova, K., Durica, M., Kovacova, M., Gregova, E., & Lazaroiu, G. (2019). Oligopolistic competition among providers in the telecommunication industry: the case of Slovakia. *Administrative Sciences*, *9*(3), 49.

[55] Matysek, A. L., & Fisher, B. S. (2016). Productivity and innovation in the mining industry. *BAEconomics Pty Ltd*: Gaber, T., El Jazouli, Y., Eldesouky, E., & Ali, A. (2021). Autonomous haulage systems in the mining industry: cybersecurity, communication and safety issues and challenges. *Electronics*, *10*(11), 1357.

[56] Akang, A. U. (2023). Regulatory compliance and access to finance: implications for business growth in developing economies. *Sciental Journal of Education Humanities and Social Sciences*, *1*(2), 8–23.

expansive operations in Australia and Brazil.[57] Additionally, these three large corporations, together with Fortescue, control nine of the world's ten largest iron ore operations.[58] These companies possess the economies of scale, infrastructure networks and marketing arrangements necessary to move hundreds of millions of tonnes of ore each year to mills across Asia and beyond.

3.5 Futures Markets, Hedging and Speculation

Commodity prices, regardless of market structure, are inherently volatile.[59] This volatility stems from a variety of factors, including the inelasticity of both supply and demand in the short term,[60] the long lead times required to develop new projects,[61] fluctuations in global economic conditions and speculative trading.[62]

To manage this volatility, mining firms and their customers often turn to commodity derivatives, a financial instrument that allows them to hedge against unfavourable price fluctuations.[63] The most common form is the futures contract: a standardised agreement to buy or sell a specific amount of a commodity at a predetermined price on an agreed future date.[64] These contracts are traded on major exchanges, including the London Metal Exchange (LME), COMEX and the Shanghai Futures Exchange (SHFE), providing liquidity and price transparency to market participants.

[57] Löf, A., & Löf, O. (2021). *Seaborne iron ore market softens.* Engineering & Mining Journal. *222*(12), 48–53. https://www.e-mj.com/features/seaborne-iron-ore-market-softens/

[58] Zadeh, J. (2025). *Energy transition's profound impact on mining economics.* Discovery Alert. https://discoveryalert.com.au/news/energy-transition-mining-renewable-clean-2025/

[59] Ezeaku, H. C., Asongu, S. A., & Nnanna, J. (2021). Volatility of international commodity prices in times of COVID-19: Effects of oil supply and global demand shocks. *The extractive industries and society*, *8*(1), 257–270.

[60] Blomberg, J., & Hellmer, S. (2000). Short-run demand and supply elasticities in the West European market for secondary aluminium. *Resources Policy*, *26*(1), 39–50.

[61] Diene, P. D., Manley, D. A. V. I. D., Olan'g, S. I. L. A. S., & Scurfield, T. H. O. M. A. S. (2022). *Triple Win: How Mining Can Benefit Africa's Citizens, Their Environment and the Energy Transition.* Natural Resource Governance Institute.

[62] Cifarelli, G., & Paladino, G. (2015). A dynamic model of hedging and speculation in the commodity futures markets. *Journal of Financial Markets*, *25*, 1–15.

[63] Cifarelli, G., & Paladino, G. (2015). A dynamic model of hedging and speculation in the commodity futures markets. *Journal of Financial Markets*, *25*, 1–15.

[64] Somanathan, T. V., Nageswaran, V. A., & Gupta, H. (2017). Futures and Forwards. In *Derivatives* (pp. 37–50). chapter, Cambridge: Cambridge University Press.

In commodity markets beyond traditional financial spheres, long-term contracts and offtake agreements[65] are crucial for determining prices, particularly in niche or greenfield commodities.[66] For example, in the lithium and cobalt sectors (critical for batteries and electric vehicles), automakers and battery manufacturers are increasingly entering into direct and indirect offtake agreements with miners.[67] These contracts secure supply in constricted markets and reduce exposure to price volatility, while providing mining companies with predictable revenue streams and cash flows that can support project financing.

Public policy and regulation also shape market structures. Governments may impose export taxes, quotas, embargoes or outright bans on strategic minerals to promote domestic value addition. Indonesia's nickel export ban, initiated in 2020, sought to encourage domestic processing and attract investment in electric vehicle supply chains.[68] While successful in some respects, such interventions can disrupt global supply and lead to temporary price distortions.[69] For example, with the coming into force of the Ghana Goldbod, the Government has strategically become the sole buyer and seller of gold within the country. This move is meant to, among others, give the government some form of power and control over the commodity traded in and outside the shores of Ghana.

[65] Distadio, L. F., & Ferguson, A. (2022). Mine offtake contracting, strategic alliances and the equity market. *Journal of Commodity Markets*, *27*, 100224

[66] Ramdoo, I., Cosbey, A., Geipel, J., & Toledano, P. (2021). *New tech, new deal: Mining policy options in the face of new technology* (Report). International Institute for Sustainable Development. https://www.iisd.org/system/files/2021-09/new-mining-technology-policy-options.pdf

[67] Bridge, G., & Faigen, E. (2022). Towards the lithium-ion battery production network: Thinking beyond mineral supply chains. *Energy Research & Social Science*, *89*, 102659.

[68] Schaap, F., & Karamira, A. (2024). *Decarbonization of the mining sector: Case studies on the role of mining in nationally determined contributions in Chile, Indonesia, and South Africa*. International Institute for Sustainable Development. https://www.igfmining.org/resource/decarbonization-of-the-mining-sector-case-studies-on-the-role-of-mining-in-nationally-determined-contributions-in-chile-indonesia-and-south-africa/

[69] Emran, S., & Tayebi, O. (2024). *"When Elephants Fight", Africa Transforms: The New Geopolitics of Critical Mineral Supply Chains and the Emergence of Local Value Chains* (Policy Paper No. PP-01/24). Policy Center for the New South.

3.6 Mining Finance and Investment Analysis

3.6.1 Resource Estimation and Economic Viability

Globally accepted standards, such as Canada's NI 43–101 and Australia's JORC Code, as well as the international CRIRSCO Template, have been developed to regulate the classification and reporting of mineral resources and reserves. These standards provide clear definitions and thresholds for confidence levels in data, distinguishing between mineral resources (which include inferred, indicated and measured categories) and mineral reserves (classified as probable or proved)—see Fig. 3.5. A mineral resource becomes a reserve only once it is deemed technically (technologically) feasible and economically viable, following the completion of detailed feasibility studies (DFS).[70]

These studies form the backbone of project evaluation and financing.[71] They progress in stages, from Preliminary Economic Assessments (PEA) and Pre-Feasibility Studies (PFS) to Detailed Feasibility Studies (DFS), the latter often referred to as Bankable Feasibility Studies (BFS).[72] Each phase provides an increasing level of detail regarding the geology, metallurgy, processing options, infrastructure requirements, capital expenditure, operating costs, market assumptions and environmental and social considerations. The resulting documents serve not only as internal decision-making tools but also as critical instruments for raising finance and securing permits.[73]

[70] Noppé, M. A. (2014). Communicating confidence in mineral resources and mineral reserves. *Journal of the Southern African Institute of Mining and Metallurgy, 114*(3), 213–213.

[71] Selim, T. H. (2004). Financial Evaluation in Project Appraisal: A Value Added Approach. In *Proceedings of the International Academy of Business and Public Administration Disciplines (IABPAD) Winter Conference. New Orleans: Louisiana, USA*; Noppé, M. A. (2014). Communicating confidence in mineral resources and mineral reserves. *Journal of the Southern African Institute of Mining and Metallurgy, 114*(3), 213–213.

[72] Kuchling, K. (2015). *4 mining study types (concept to feasibility)*. KJ Kuchling Consulting Ltd. https://kuchling.com/four-study-stages-concept-to-feasibility-which-should-we-do/; Noppé, M. A. (2014). Communicating confidence in mineral resources and mineral reserves. *Journal of the Southern African Institute of Mining and Metallurgy, 114*(3), 213–213.

[73] Selim, T. H. (2004). Financial Evaluation in Project Appraisal: A Value Added Approach. In *Proceedings of the International Academy of Business and Public Administration Disciplines (IABPAD) Winter Conference. New Orleans: Louisiana, USA;* See the example of the Ewoyaa Definitive Feasibility Study in Ghana here at https://static1.squarespace.com/static/61711d27ed0db12cacbcfb5a/t/649d3f851c02121fcf7bffcd/1688027016483/2023.06.29+-+Project+Update+-+Ewoyaa+Definitive+Feasibility+Study+%28ASX%29.pdf

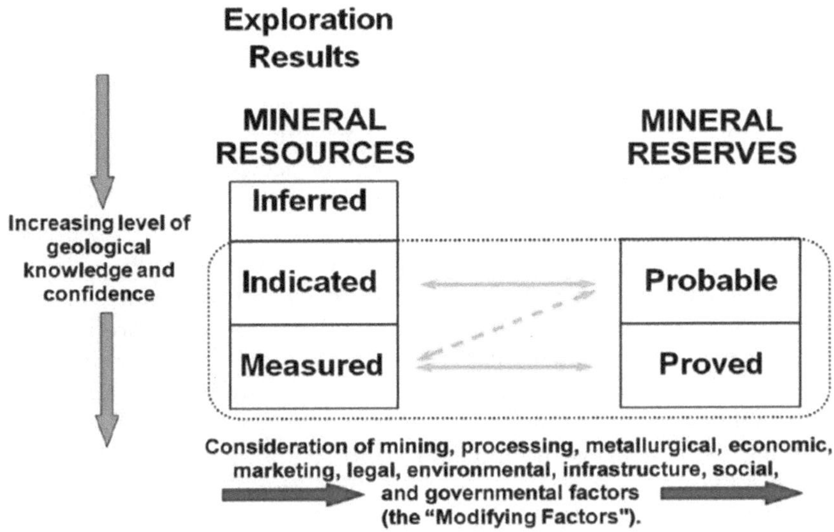

Fig. 3.5 General relationship between exploration results, mineral resources and mineral reserves. (Source: CRIRSCO (2019))

3.6.2 Valuation Fundamentals and Approaches

Once a project's reserves have been confirmed, it must be valued. This is a complex exercise that balances financial forecasting, market expectations and project-specific risks.[74] The most widely used approach is the Discounted Cash Flow (DCF) methodology,[75] which involves projecting the revenues and costs associated with the mine over its life and then discounting those future cash flows back to present value using a chosen rate. Various investment valuation tools are used, including:

- Net Present Value (NPV)
- Adjusted present value (APV)
- Payback period/discounted payback

[74] Rudenno, V. (2012). *The mining valuation handbook 4e: mining and energy valuation for investors and management*. John Wiley & Sons; Rothman, T., & Rothman, T. (2020). Understanding Financial Valuations: Foundations and Basic Traditional Techniques. *Valuations of Early-Stage Companies and Disruptive Technologies: How to Value Life Science, Cybersecurity and ICT Start-ups, and their Technologies*, 7–20.

[75] Einstein, B. (2025). *Discounted Cash Flow (DCF) Formula: What It Is & How to Use It*. HBS Online. https://online.hbs.edu/blog/post/discounted-cash-flow

- Profitability Index (PI)
- Internal Rate of Return (IRR)

NPV is a mathematical calculation involving net cash flow at a particular present time 't' at discount rate at the same time. The basic rule is that if NPV > 0, the project is viable and should be accepted; If NPV < 0, the project is not viable and should be rejected. In tandem with NPV, the Internal Rate of Return (IRR), which is the discount rate at which NPV equals zero, is used to gauge the efficiency of an investment. Projects with higher internal rates of return (IRR) are typically more attractive, especially to investors seeking quick payback and strong capital productivity. Additionally, the Payback Period or the time it takes to recover the initial investment, is a common decision-making metric, especially for risk-averse stakeholders.[76]

While DCF remains the industry standard for valuing producing or near-producing assets, it is not without limitations. Among others, the method is highly sensitive to input assumptions[77] such as metal prices, operating costs, discount rates, and project timelines.[78] Moreover, DCF assumes a single forecasted future, which rarely captures the uncertainty inherent in mining operations.[79] Due to this, experienced practitioners pair DCF with another layer of sensitivity analysis,[80] which tests how valuation outcomes respond to changes in key parameters.

3.6.3 Sensitivity Analysis, Scenario Planning, and Dynamic Valuation Techniques

Specific to the sensitivity modelling, variables such as commodity price, capital costs, operating expenses and exchange rates are adjusted one after the other to determine which factors have the greatest influence on project

[76] Ediriweera, A., & Wiewiora, A. (2021). Barriers and enablers of technology adoption in the mining industry. *Resources Policy*, *73*, 102188.

[77] Einstein, B. (2025). *Discounted Cash Flow (DCF) Formula: What It Is & How to Use It.* HBS Online. https://online.hbs.edu/blog/post/discounted-cash-flow

[78] Kumar, V., & Sharma, R. (2023). Valuation of Damages in Cases of Unlawful Expropriation of Data Assets: Pitfalls of the DCF Method. *Available at SSRN 4890685.*

[79] Lilford, E., & Haque, A. (2023). Limitations and Issues With Using The DCF NPV Method For Valuations Of Mine Closure And Post Mine Closure. *CRC TiME, Perth.*

[80] Barry, S. (2022). *Sensitivity analysis in project evaluation.* SRK Consulting (SRK News Issue #64). https://www.srk.com/en/publications/sensitivity-analysis-in-project-evaluation

metrics. Tools such as spider diagrams and tornado charts help visualise these dependencies, highlighting where management should focus risk mitigation efforts. When two or more variables are considered together, scenario analysis becomes a valuable tool.[81]

In addition to these traditional approaches, the mining industry has increasingly turned to more dynamic valuation techniques such as Real Options Analysis (ROA).[82] This method treats investment decisions as options that can be deferred, expanded, contracted, or even abandoned, depending on how future conditions evolve. ROA is particularly useful for long-life assets with significant optionality, such as staged expansions or alternative development routes, and for commodities subject to high volatility, like lithium or cobalt.[83] Real options can often justify higher valuations than rigid DCF models allow.[84]

3.6.4 Financial Structuring and Project Financing

In addition to valuation, mining projects must also be structured financially to attract the right mix of capital at each stage of development. Early-stage exploration is typically funded through equity,[85] either via private placements or listings on exchanges such as the Toronto Stock Exchange (TSX), the Australian Securities Exchange (ASX) or London's Alternative Investment Market (AIM). As the project progresses and risk

[81] Jackson, S., Vann, J. E., Coward, S., & Moayer, S. (2014). Scenario-based project evaluation—Full mineral value chain stochastic simulation to evaluate development and operational alternatives. In *Proceedings of the International Mining Geology Conference, Adelaide, Australia* (pp. 18–20).

[82] Mun, J. (2012). *Real options analysis: Tools and techniques for valuing strategic investments and decisions* (Vol. 320). John Wiley & Sons.

[83] UN DESA. (2025). *Harnessing the potential of critical minerals for sustainable development.* https://desapublications.un.org/sites/default/files/publications/2025-01/WESP%202025_Harnessing%20the%20Potential%20of%20Critical%20Minerals%20for%20Sustainable%20Development_WEB.pdf

[84] Savolainen, J. (2016). Real options in metal mining project valuation: Review of literature. *Resources Policy, 50*, 49–65; Dzyuma, U. (2012). Real options compared to traditional company valuation methods: possibilities and constraints in their use. *Finansowy Kwartalnik Internetowy e-Finanse, 8*(2), 51–68.

[85] Fraser, J., Kunz, N. C., & Batdorj, B. (2019). Can mineral exploration projects create and share value with communities? A case study from Mongolia. *Resources Policy, 63*, 101455; Peña Alfaro, I. (2025). *Equity financing: Benefits and risks for startups.* Rho. https://www.rho.co/blog/equity-financing-for-startups

is mitigated, debt becomes a more viable option, particularly during the construction and production phases.

The choice between equity and debt is strategic.[86] Equity is flexible and absorbs early-stage risk,[87] but it dilutes ownership and is expensive in terms of return expectations. Debt offers a lower cost and avoids dilution, but it imposes fixed repayment obligations and may come with restrictive agreements. Mining companies often balance these considerations carefully, employing a mix of both. Projects that rely too heavily on debt may appear more profitable in good times due to lower cost of capital.[88] However, they become vulnerable during downturns when revenues fall short of debt service obligations. Conversely, over-reliance on equity avoids cash flow pressure but reduces return on equity and can depress share price performance. The Weighted Average Cost of Capital (WACC), which blends debt and equity costs, is a central metric in determining valuation. Skilled financial structuring aims to optimise this balance to deliver the highest NPV (or other metrics) without overexposing the company to financial stress.

Furthermore, ESG factors have become embedded in investment analysis. Institutional investors now demand compliance with international standards such as the Equator Principles, IRMA (Initiative for Responsible Mining Assurance), ICMM (2020) Mining Principles and OECD (2023) Due Diligence Guidance for Responsible Mineral Supply Chains, among others. Projects with poor environmental or social performance increasingly face higher cost of capital, reduced funding options or complete investor exclusion. Some lenders and rating agencies now incorporate carbon pricing[89] into project assessments, applying an internal cost per tonne of carbon dioxide equivalent emitted. Projects that do not have a decarbonisation plan or that rely heavily on fossil fuels face elevated risk

[86] Wang, C. (2024). Debt vs. equity: Analyzing the optimal capital structure for businesses. *Business and Economics Journal*, *15*(3), 1–2. https://www.hilarispublisher.com/open-access/debt-vs-equity-analyzing-the-optimal-capital-structure-for-businesses-107805.html

[87] Griffin, J. G. (2022). Equity: Balancing certainty and flexibility to secure justice. *Fields: Journal of Huddersfield Student Research*, *8*(1), 1–21.

[88] Arhinful, R., & Radmehr, M. (2023). The Impact of Financial Leverage on the Financial Performance of the Firms Listed on the Tokyo Stock Exchange. *SAGE Open*, *13*(4).

[89] Downey, A. (2024). *What's the role of credit rating agencies in carbon markets?* Sylvera. https://www.sylvera.com/blog/what-is-a-carbon-credit-agency

premiums.[90] On the other hand, those incorporating renewables, community partnerships and transparent governance are increasingly able to access sustainability-linked loans and ESG-focused equity funds.

The drive towards sustainability has also created market differentiation. Buyers are beginning to offer premiums for low-carbon or responsibly sourced metals. The emergence of products like 'green aluminium' or 'ESG-certified copper' reflects growing demand for traceable, ethically produced materials. Blockchain-based platforms and digital certification schemes are enabling the verification of ESG claims, providing a competitive edge to early adopters. These trends are influencing the way mining projects are evaluated by the financial community. Traditional Discounted Cash Flow (DCF) models are being supplemented by Integrated Risk Modelling (IRM),[91] which incorporates operational, environmental, political and ESG risks into probabilistic frameworks.

3.6.5 Economic Impacts of Taxation and Incentive Schemes

Extractive fiscal regimes encompass various instruments, including corporate income taxes, royalties, import and export duties, withholding taxes, and occasionally, state equity participation or production-sharing agreements. Their primary objective is to ensure that host governments derive fair economic returns from their mineral wealth while also remaining attractive to investors.[92]

Royalties are mechanisms by which governments receive payment for allowing the extraction of publicly owned resources.[93] Ad valorem royalties, calculated as a percentage of gross production, are straightforward and ensure stable government revenue even when profits are thin. However, because they are levied regardless of profitability, they can

[90] Calcaterra, M., et al. (2024). Reducing the Cost of Capital to Finance the Energy Transition in Developing Countries. Nature Energy 9, 1241. https://doi.org/10.1038/s41560-024-01606-7. Accessed 27 June 2026.

[91] Barney, N., & Cha, W. (2023). *Integrated risk management (IRM)*. TechTarget. https://www.techtarget.com/searchsecurity/definition/integrated-risk-management-IRM

[92] Acheampong, Theophilus, and Abdallah Ali-Nakyea. "Competitiveness of Ghana's Upstream Petroleum Fiscal Regime: Fit for Purpose?." *Petroleum Resource Management in Africa: Lessons from Ten Years of Oil and Gas Production in Ghana*. Cham: Springer International Publishing, 2022. 157–205.

[93] Otto, J. (2006). *Mining royalties: A global study of their impact on investors, government, and civil society*. World Bank Publications

increase financial pressure during periods of price downturns or high capital expenditures. Unit-based royalties, which charge a fixed amount per tonne or ounce of production, are simpler but less responsive to market conditions. Profit-Based Royalties (PBR), such as Net Smelter Returns (NSR) or Net Profit Interests (NPI), vary with project economics and are arguably fairer to investors; however, they require greater transparency and administrative capacity to audit costs.

3.6.5.1 Incentive Structures and Investment Attraction

To encourage exploration and development, many governments offer fiscal incentives, especially in emerging mining jurisdictions. These can include tax holidays, accelerated depreciation of capital expenditure, loss carry-forward allowances, and even co-investment in infrastructure. For example, Namibia and Botswana have used such incentives to attract diamond and uranium investments, offering favourable tax terms in exchange for local employment commitments and downstream beneficiation.[94]

However, incentives must be carefully balanced. Overly generous fiscal terms deprive governments of revenue, while poorly designed fiscal regimes often fail to capture upside in commodity booms. Increasingly, countries are incorporating progressive royalty systems where the rate increases as prices rise, ensuring that windfalls are shared without undermining base-case economics. Bouterige et al. find that mining royalty rates are rising in Africa, with countries like Zambia and Mauritania introducing variable rates based on commodity prices.[95]

In Latin America, major producers like Chile and Peru have also adopted progressive royalty systems linked to companies' profit margins. For Indonesia, proposals are being considered to reform the nickel royalty system by replacing the current flat 10% rate with a progressive structure, with royalties ranging from 14% to 19%, depending on benchmark nickel prices.[96] The integration of Production Sharing Contracts in mining

[94] Mannathoko, I.M. (2019). Governance in the Mineral Dependent Economy: The Case of Botswana. In: Elhiraika, A., Ibrahim, G., Davis, W. (eds) Governance for Structural Transformation in Africa. Palgrave Macmillan, Cham. https://doi.org/10.1007/978-3-030-03964-6_5

[95] Bouterige, Y., Amedanou, Y., & Laporte, B. (2023). Mining tax regimes and rent sharing: Comparative analysis Africa—Latin America, the case of copper. *Revue de Droit Fiscal, (11–12)*, LexisNexis.

[96] Discovery Alert. (2025). *Indonesia's mining royalty revolution: Strategic economic transformation in 2025.* https://discoveryalert.com.au/news/indonesia-mining-royalties-boosting-2025-economy/

arrangements, much like the oil and gas sector, is also gaining traction in many resource-rich developing countries.[97] This is partly due to the tendency for such contracts to drive and promote investments by aligning the interests of governments and investors.[98] Moreover, the fiscal and tax provisions, stability clauses and legal provisions of PSCs are crucial for managing risks, controls and profit-sharing across various phases of the mining project cycle.[99] Specific to Africa, Margolin affirms that PSCs are predominantly being infused into African mining legislations, though at a steady pace.[100]

Each country adopts a fiscal system based on its political economy, mineral endowment and administrative capabilities. For instance, Chile, the world's largest copper producer,[101] has historically employed a relatively stable tax and royalty regime, which has supported long-term investment while allowing the state to capture significant rents through CODELCO, its state-owned mining company. However, there are ongoing debates over royalty reform and constitutional reforms.[102]

[97] International Institute for Sustainable Development (IISD) and African Tax Administration Forum (ATAF), Designing and Implementing Production Sharing Contracts for the Mining Sector (IGF and ATAF, November 2022). https://www.igfmining.org/wp-content/uploads/2022/11/designing-and-implementing-production-sharing-contracts-for-themining-sector.pdf. Accessed 27 June 2026.

[98] Feng, Z., Zhang, S., & Gao, Y. (2014). On Oil Investment and Production: A Comparison of Production Sharing Contracts and Buyback Contracts. *Energy Economics, 42,* 395–402.

[99] Yan, H., & Zhou, P. (2024). Terms of Production Sharing Contracts in International Oil and Gas Cooperation and China's Optimization Path' (2024). *Journal of Law and Judicial System 7*(1), 27–33.

[100] Margolin, L. (2025). Will production sharing contracts be the cure for mining? *Africa Business Plus.* https://www.africabusinessplus.com/en/823877/will-production-sharing-contracts-be-the-cure-for-mining/

[101] Meller, P., & Simpasa, A. M. (2011). Role of copper in the Chilean & Zambian economies: Main economic and policy issues (GDN Working Paper No. 43). *Global Development Network.*

[102] Viale, C. (2020). Reforming Colombia's extractive sector royalty distributions: Key changes and coronavirus. *Natural Resource Governance Institute.* https://resourcegovernance.org/articles/reforming-colombias-extractive-sector-royalty-distributions-key-changes-and-coronavirus; Zhou, Z., Liu, J., & Zhang, H. (2023). Does the reform of China's mineral royalty policies exert economic pressure on mining companies?. *The Extractive Industries and Society, 15,* 101325.

3.7 Impact of New Technologies on Valuation Assumptions

Automation has emerged as a game-changer in large-scale mining operations by reducing labour costs and improving safety.[103] Similarly, remote operations centres enable centralised control over geographically dispersed mines, making it possible to manage production and respond to anomalies in real time. Artificial Intelligence (AI) and machine learning technologies are enhancing exploration success rates by analysing vast datasets, including satellite imagery and geochemical assays, to identify promising mineral targets.[104] These tools reduce the time and cost of discovery while improving the accuracy of geological models.

Ore-sorting technologies are also enabling precision mining, allowing companies to process only high-value material and discard waste early in the production chain.[105] This not only improves efficiency but also significantly reduces energy consumption, water usage and tailings generation.

Renewable energy has also made its mark. Today, many operators are integrating solar, wind and battery systems into their energy mix. In countries like Chile, where high solar irradiance coincides with substantial copper reserves,[106] integrating renewable energy sources is increasingly cost-effective.[107] Mines in South Africa, Australia and Peru[108] are following suit, reducing both operating costs and carbon footprints. These technological shifts have significant implications for valuation.[109] Traditional

[103] Leung, R., Hill, A. J., & Melkumyan, A. (2023). Automation and Artificial Intelligence Technology in Surface Mining: A Brief Introduction to Open-Pit Operations in the Pilbara. *IEEE Robotics & Automation Magazine*.

[104] Davies, R. S., Trott, M., Georgi, J., & Farrar, A. (2025). Artificial intelligence and machine learning to enhance critical mineral deposit discovery. *Geosystems and Geoenvironment*, *4*(2), 100361.

[105] Normanyo, E., Ofosu, R. A., & Acquah, D. S. (2022). On sensor-based ore sorting. *Ghana Mining Journal*, *22*(1), 56–61.

[106] Meller, P., & Simpasa, A. M. (2011). Role of copper in the Chilean & Zambian economies: Main economic and policy issues (GDN Working Paper No. 43). *Global Development Network*.

[107] Gonzales, L. E., Ito, K., & Reguant, M. (2024). Expanding renewable energy: Lessons from Chile. *Microeconomic Insights*. https://microeconomicinsights.org/expanding-renewable-energy-lessons-from-chile/

[108] Gold Fields Limited. (2024). *Climate Change and Environment Report 2024*. https://www.goldfields.com/reports/annual-report-2023/environmental-stewardship.php

[109] Flores-Castañeda, R. O., Olaya-Cotera, S., López-Porras, M., Tarmeño-Juscamaita, E., & Iparraguirre-Villanueva, O. (2024). Technological advances and trends in the mining industry: a systematic review. *Mineral Economics*, 1–16.

financial models are also being adjusted to account for lower operating costs, higher recovery rates and reduced environmental risk due to innovation. Investors are also rewarding companies that adopt technology not just for economic gains but to meet ESG benchmarks.[110]

3.8 Conclusion

This chapter has explored the foundations of mining economics, ranging from the principles of scarcity and cost dynamics to market structures, financial strategy, taxation, innovation and sustainability. It is apparent that mining is not merely a matter of extracting value from the ground; it is about navigating uncertainty, balancing interests and building the conditions for shared and sustained prosperity.

At one end of the spectrum lie the mineral deposits, which are finite, often remote, and increasingly complex to extract. At the other end are the markets and economies that depend on these resources for energy, technology, infrastructure and industrial growth. Between these poles stand the economic models, investment frameworks, regulatory systems and social contracts that govern how resources are developed, distributed and valued.

A recurring theme throughout this chapter is the importance of economic viability as a moving target. What is profitable today may not be tomorrow as metal prices shift, regulations evolve or operating costs rise. Conversely, deposits previously deemed subeconomic may become viable with the introduction of new technologies, rising demand and/or supportive fiscal regimes. Mining economics, therefore, is not static but dynamic, scenario-driven, and increasingly linked with broader global trends such as climate action, digitalisation and supply chain reconfiguration. With capital requirements often in the hundreds of millions or billions of dollars, funding decisions require robust economic evaluation and creative structuring. Discounted cash flow models, sensitivity analyses and real options valuation remain core tools. However, they are now being complemented by ESG screens, carbon pricing forecasts and integrated risk models that reflect a more holistic view of value and vulnerability.

[110] Bajpai, M., Tripathi, A., & Prasad, N. (2023). Comparative analysis of traditional and modern financial models in investment decision making. *ShodhKosh: Journal of Visual and Performing Arts, 4*(2). https://doi.org/10.29121/shodhkosh.v4.i2.2023.2219

CHAPTER 4

Geopolitics of Africa's Critical Minerals Industry

Abstract Achieving net-zero energy requires major changes in energy production, storage and use, including expanded renewable technologies. Geopolitical tensions, especially following the Russia–Ukraine war, have intensified concerns over energy and critical mineral supply chains. Critical minerals are increasingly viewed as the 'new oil' due to their strategic and geopolitical importance. The World Bank estimates that three billion tonnes of critical minerals will be needed to decarbonise the global energy system by 2050. Demand for critical minerals is expected to triple by 2040, driven by the green and digital transitions. The classification of critical minerals varies by country and over time, based on economic, technological and security needs. Supply chain risks are heightened by geographic concentration, with countries like China dominating mineral refining capacity. Governments are adopting strategic policies, such as the U.S. Inflation Reduction Act and the EU Green Deal, to secure mineral supplies. Africa holds about 30% of global critical mineral reserves, positioning it as a vital player in the energy transition. Effective governance, transparency, and sustainability are essential for Africa to translate mineral wealth into long-term development gains. African governments are seeking to move beyond raw mineral exports by promoting local value

This chapter was written with contribution from Isaac Owusu Asare.

© The Author(s), under exclusive license to Springer Nature Switzerland AG 2026

R. J. Heffron et al., *Revolution in the Global Critical Minerals Sector*, Just Transitions,
https://doi.org/10.1007/978-3-032-26366-7_4

addition and sustainable economic development. Continental and national initiatives, such as the Africa Mining Vision and export restrictions, aim to ensure mineral wealth benefits African citizens. Countries like the DRC, Zimbabwe, Namibia and Ghana are encouraging local processing through bans or limits on raw mineral exports. Global powers are competing for influence in Africa's mineral supply chains, with efforts to counter China's dominance and form new alliances. Public–private partnerships, such as the Lobito Corridor project, are emerging as key mechanisms to finance mining and infrastructure development.

Keywords Critical minerals • Geopolitics of minerals • Resource security • Supply chain resilience • Strategic minerals • Export controls • International trade • Energy transition • Global power dynamics • China mineral dominance • US mineral strategy • European Union mineral policy • Africa mineral geopolitics

4.1 Introduction

Switching to net-zero energy sources requires radically different means of producing, using and storing energy—upscaling renewables, including solar PV, wind, hydrogen and new mobility solutions. Nevertheless, the aftershocks of the Russia-Ukraine war and increased geopolitical tensions have compelled many economies to implement various policies, laws and regulations to mitigate strategic supply chain risks for energy resources and the critical raw materials required for the energy transition.[1] Critical minerals have been characterised as the 'new oil' of the twenty-first century, with inherent supply risks and geopolitical significance. According to the World Bank, 3 billion tonnes of critical minerals are needed to decarbonise the global energy system by 2050.[2] In the wake of the ongoing

[1] Catalán M, Natalucci F, Qureshi MS and Tsuruga T, 'Geopolitics and Fragmentation Emerge as Serious Financial Stability Threats' (IMF, 2023). https://www.imf.org/en/Blogs/Articles/2023/04/05/geopolitics-and-fragmentation-emerge-as-serious-financial-stability-threats accessed 23 June 2025; International Energy Agency, Global Critical Minerals Outlook Opportunities and challenges for resilient supply chains (2024) https://unece.org/sites/default/files/2025-03/02.%20Tae-Yoon_Kim.pdf accessed 1 May 2025.

[2] Hund K, La Porta D, Thao F, Laing T and Drexhage J, Minerals for Climate Action: The Mineral Intensity of the Clean Energy Transition (World Bank Group, Washington DC, 2020) http://documents.worldbank.org/curated/en/099052423172525564; World Bank Group, 'Mineral Production to Soar as Demand for Clean Energy Increases' (2020)

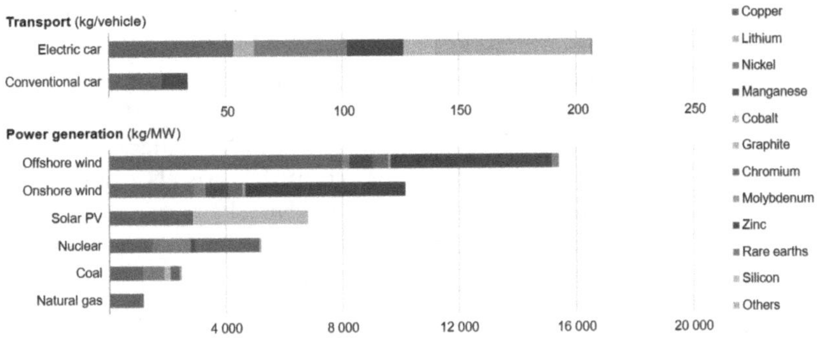

IEA. All rights reserved.

Notes: kg = kilogramme; MW = megawatt. Steel and aluminium not included. See Chapter 1 and Annex for details on the assumptions and methodologies.

Fig. 4.1 Minerals used in selected clean energy technologies. (Source: IEA, 2020, p.6)

'twin transitions' of the green and digital industries, the demand for these critical minerals is expected to triple by 2040.[3] The term 'critical minerals' is often used interchangeably with the terms 'strategic minerals' and 'energy transition minerals'.[4]

While classifying a 'critical mineral' is subjective, there are commonly accepted criteria for making that determination. Traditionally, these criteria have been associated with the role these minerals play in the economy, military, information and communication technologies, and other strategic industries (Fig. 4.1). A key factor is that these criteria vary with a country's needs and interests at any given time. For example, table salt was once considered a critical mineral.[5] Geopolitical considerations have been prominent in the extant literature. These factors are a critical source of supply chain risks associated with these critical minerals.

https://www.worldbank.org/en/news/press-release/2020/05/11/mineral-production-to-soar-as-demand-for-clean-energy-increases accessed 23 June 2025.

[3] International Energy Agency, Global Critical Minerals Outlook Opportunities and challenges for resilient supply chains (2024) https://unece.org/sites/default/files/2025-03/02.%20Tae-Yoon_Kim.pdf accessed 1 May 2025.

[4] International Energy Agency, The Role of Critical Minerals in Clean Energy Transitions, World Energy Outlook Special Report (2022) https://iea.blob.core.windows.net/assets/ffd2a83b-8c30-4e9d-980a-52b6d9a86fdc/TheRoleofCriticalMineralsinCleanEnergyTransitions.pdf accessed 1 May 2025.

[5] American Geosciences Institute, Critical Minerals (2025) https://profession.americangeosciences.org/society/intersections/critical-minerals/ accessed 1 May 2025.

This situation gives countries significant mineral concentrations and considerable economic and political power, which can sometimes be described as 'unfair'. For example, nearly 50% of the market value from CRM refining is expected to be concentrated in China by 2030.[6] Thus, the quest to control the global supply chain of these minerals is fierce. Given the national security implications, countries are also developing comprehensive strategic plans to secure their fair share of supplies and locally added value. For instance, in August 2022, the U.S. government enacted the Inflation Reduction Act (IRA), which allocated approximately US$400 million from tax breaks to federal spending on clean energy, thereby ramping up the demand for critical minerals.[7] Additionally, in the European Union, the EU Green Deal policy aims for at least a 55% reduction in greenhouse gas emissions (GHGs) by 2030.[8] The rising demand for these minerals places immense pressure on extraction, production and refining processes, presenting threats and opportunities for governments to navigate.

Africa stands at a crucial crossroads where it must fully leverage this significant demand for critical minerals. Given its challenging history with ineffective resource management, particularly regarding oil and gas resources,[9] analysts assert, "Africa must not let this opportunity pass it by".[10] In addition to the potential economic benefits, the continent stands to gain global political influence and power. Africa can also assert its voice in international affairs, especially in the climate change and net-zero

[6] International Energy Agency, Global Critical Minerals Outlook Opportunities and challenges for resilient supply chains (2024) https://unece.org/sites/default/files/2025-03/02.%20Tae-Yoon_Kim.pdf accessed 1 May 2025.

[7] Trost JN and Dunn JB, 'Assessing the feasibility of the Inflation Reduction Act's EV critical mineral targets' (2023) 6 Nature Sustainability 639; Church S, Laske N, Leschiner D, Simmons N, Verhalen S and Willis M, 'The Impact of the US Inflation Reduction Act on Global Clean Energy Supply Chains' (American University Diplomacy Lab, 2023); Li J, Wang D and Qin M, 'Understanding the role of mineral resources and inflation in promoting sustainable development under the inflation reduction act' (2024) 90 Resources Policy 104757.

[8] European Commission, The European Green Deal (2019) https://commission.europa.eu/strategy-and-policy/priorities-2019-2024/european-green-deal_en accessed 4 June 2025.

[9] World Bank, Avoiding the Resource Curse (2017) https://documents1.worldbank.org/curated/en/959621468227948158/pdf/397340ENGLISH0KH0Oil1and1Gas0301PUBLIC1.pdf accessed 4 May 2025.

[10] Mo Ibrahim Foundation, Africa's Critical Minerals (2022) https://mo.ibrahim.foundation/sites/default/files/2022-11/minerals-resource-governance.pdf accessed 2 May 2025.

debates from which it has long been marginalised.[11] The pursuit of a low global carbon future cannot be realised without Africa's critical mineral reserves, which are estimated to account for 30% of the world's total.[12] The strategic importance of these minerals, especially given the fierce competition between the U.S. and China, is positioning Africa as a key player in the global supply chain.[13] For the continent, prudent resource governance, including contract transparency, accountability and environmental sustainability, will be essential in leveraging this newfound resource potential into development outcomes.

4.2 Mapping Africa's Critical Mineral Resource Endowments

Africa has an abundance of these CRMs needed for the energy transition. Major CRM resource-endowed countries include Zambia, the Democratic Republic of the Congo (DRC), Gabon, Zimbabwe, South Africa, Tanzania, Guinea and Ghana (Table 4.1). For graphite, major producers on the continent include Mozambique (7%) and Madagascar (2%). In 2021, Africa accounted for approximately 9% of the global graphite flake supply.[14] Pipeline projects include production from Namibia, Tanzania and Guinea, as well as the expansion of existing supplies from Mozambique and Madagascar.

Regarding copper, Zambia and the DRC collectively account for approximately 5.6% of global copper reserves, 13.5% of global mine production, and 7.9% of refinery production, respectively. These two countries collectively account for more than 70% of the continent's output.[15]

[11] ibid.

[12] Atlantic Council, Critical Minerals Task Force (2024) https://www.atlanticcouncil.org/programs/africa-center/critical-minerals-task-force/ accessed 4 May 2025.

[13] Acheampong T, 'From mines to markets: How Africa and Europe can become green industry partners of choice' (2024a) https://ecfr.eu/publication/from-mines-to-markets-how-africa-and-europe-can-become-green-industry-partners-of-choice/ accessed 1 June 2025; Acheampong T, 'Rules of entry: Cooperative pathways for Europe and Africa on critical minerals' (2024b) https://ecfr.eu/publication/rules-of-entry-cooperative-pathways-for-europe-and-africa-on-critical-minerals/ accessed 2 June 2025.

[14] Taylor G, 'Africa's Growing Graphite Mining Potential' (2022) S&P Global Market Intelligence https://www.spglobal.com/market-intelligence/en/news-insights/research/africa-s-growing-graphite-mining-potential accessed 23 June 2025.

[15] USGS, Copper (2023a) https://pubs.usgs.gov/periodicals/mcs2023/mcs2023-copper.pdf accessed 23 June 2025.

Table 4.1 Example of Africa's share of green minerals in world production and reserves 2023 (estimate)

Mineral	*World production*	*World reserve*	*Top African countries*	*Production (% of world)*	*Reserves (% of world)*
Aluminium (smelter prod)	70 Mt		South Africa	720 kt (1%)	
			Mozambique	314 kt (0.5%)	
			Egypt	320 kt (0.5%)	
Alumina	140 Mt		Guinea	330 kt (0.3%)	
Bauxite	400 Mt	30 Gt	Guinea	97 Mt. (24.3%)	7.4 Gt (23.8%)
Chromium	41 Mt	560 Mt	South Africa	18 Mt. (43.9%)	200 Mt. (35.7%)
			Zimbabwe	1.4 Mt. (3.4%)	
Cobalt	230 Mt	11 Mt	DRC	170 kt (73.9%)	6 Mt. (54%)
Copper mine production	22 Mt	1 Gt	DRC	2.5 Mt. (11.3%)	80 Mt. (8%)
			Zambia	760 kt (3.4%)	21 Mt. (2.1%)
Copper refinery	27 Mt		DRC	1.9 Mt. (7%)	
			Zambia	380 kt (1.3%)	
Graphite - natural	1.6 Mt	280 Mt	Mozambique	96 kt (6%)	25 Mt. (8.9%)
			Madagascar	100 kt (6.2%)	24 Mt. (8.5%)
			Tanzania	6 kt (0.3%)	18 Mt. (5.5%)
Iron & steel (raw steel)	1.9 Gt				
Iron ore (Fe cont.)	1.5 Gt	87 Gt	South Africa	39 Mt. (2.6%)	600 Mt. (0.6%)
			Mauritania	8.1 Mt. (0.5%)	N/A
Lithium	180 Kt	28 Mt	Zimbabwe	3.4 kt (1.8%)	310 kt (1.1%)
Manganese	20 Mt	1.9 Gt	South Africa	7.2 Mt. (36%)	600 Mt. (31.5%)
			Gabon	4.6 Mt. (23%)	61 Mt. (3.2%)
			Ghana	840 kt (4.2%)	13 Mt. (0.7%)

Source: Africa Green Minerals Strategy, 2025: p.16

Other copper producers include South Africa, Morocco, Mauritania and Zimbabwe.[16] On the other hand, major bauxite producers include Guinea, Sierra Leone, Mozambique and Ghana.[17] Countries with aluminium smelters in Africa include Ghana, South Africa, Cameroon, Mozambique and Egypt.[18]

Gabon, Côte d'Ivoire, Ghana and South Africa are the primary manganese producers in Africa. These countries collectively accounted for an estimated 66% of mine production in 2022.[19] For nickel, South Africa, Zimbabwe, Tanzania, Zambia and Madagascar have significant nickel deposits on the continent.[20] The DRC is the world's largest producer of cobalt, accounting for approximately 70% of global production.[21] Other producers include South Africa and Zambia. Finally, countries on the continent with identified lithium resources and some projects at the development stage, likely to come onstream in the next five years, include Ghana, Namibia, Zimbabwe, the DRC, Mali and Ethiopia.[22]

[16] Sasu DD, 'Africa: copper mine production by country 2021' (2025) Statista https://www.statista.com/statistics/1049848/african-copper-mine production by country/ accessed 23 June 2025.

[17] Zainudeen NM, Mohammed L, Nyamful A, Adotey D and Osae SK, 'A comparative review of the mineralogical and chemical composition of African major bauxite deposits' (2023) Heliyon 9(8); Shankar B, 'Profiling the top four bauxite exporting countries in Africa' (2021) https://www.nsenergybusiness.com/news/bauxite-exporting-countries-africa/ accessed 23 June 2025.

[18] Husband C, McMahon G and Van Der Veen P, 'The Aluminum industry in West and Central Africa: Lessons learned and prospects for the future' (2009); Acheampong T and Tyce M, 'Navigating the energy transition and industrial decarbonisation: Ghana's latest bid to develop an integrated bauxite-to-aluminium industry' (2024) Energy Research & Social Science 107, 103337.

[19] USGS, Manganese (2023b) https://pubs.usgs.gov/periodicals/mcs2023/mcs2023-manganese.pdf accessed 23 June 2025.

[20] Energy Capital & Power, 'Top 10 Lithium Mines in Africa' (2023) https://energycapitalpower.com/top-10-lithium-mines-in-africa accessed 23 June 2025.

[21] USGS, Cobalt (2023c) https://pubs.usgs.gov/periodicals/mcs2023/mcs2023-cobalt.pdf accessed 23 June 2025.

[22] Energy Capital & Power, 'Top 10 Lithium Mines in Africa' (2023) https://energycapitalpower.com/top-10-lithium-mines-in-africa accessed 23 June 2025; Global Witness, 'The lithium rush in Africa' (2023) https://globalwitness.org/en/campaigns/transition-minerals/a-rush-for-lithium-in-africa-risks-fuelling-corruption-and-failing-citizens accessed 23 June 2025.

4.3 International Interests and Influence: China's Strategic Positioning in the African Mining Sectors

The rising demand for, and limited current supplies of, critical minerals have sparked significant interest in Africa in securing the raw materials needed for manufacturing various energy-efficient products. China currently dominates the midstream and downstream sectors of the global critical minerals supply chain. In the midstream, it leads in the refining and processing of these minerals, capturing about 50% of the sector's market value.[23] Where China lags is in its dominant control of upstream production in some minerals (Fig. 4.2). Amid growing concerns about the vast influence and dependence on China in the quest for global decarbonisation, and with strong strategic efforts from the U.S. and Europe to reduce the world's reliance on China for critical minerals, the East Asian giant has also taken steps to maintain its dominance in the global supply chain through strategic positioning in Africa's mining sector. To this end, China has a significant interest in mineral-rich African countries, making substantial financial investments in exploration and extraction activities. The Chinese approach also encompasses a range of investments that span the entire supply chain, thereby solidifying its presence in Africa.[24] Beijing remains highly active on the continent, with substantial financial and capital investments.

Western governments, particularly the United States and the European Union, are also establishing their supply chains to counter Chinese dominance, including in Africa. These initiatives, outlined in their strategy documents, aim to reduce dependence on China. In August 2022, the United States of America launched the Inflation Reduction Act (IRA), offering new spending and tax incentives to boost clean energy, among other initiatives.[25] Approximately US$370 billion is intended to be disbursed for energy security and accelerating clean energy transition initiatives,

[23] International Energy Agency, Global Critical Minerals Outlook Opportunities and challenges for resilient supply chains (2024) https://unece.org/sites/default/files/2025-03/02.%20Tae-Yoon_Kim.pdf accessed 1 May 2025.

[24] Acheampong T and Tyce M, 'Navigating the energy transition and industrial decarbonisation: Ghana's latest bid to develop an integrated bauxite-to-aluminium industry' (2024) Energy Research & Social Science 107103337.

[25] Trost JN and Dunn JB, 'Assessing the feasibility of the Inflation Reduction Act's EV critical mineral targets' (2023) 6 Nature Sustainability 639.

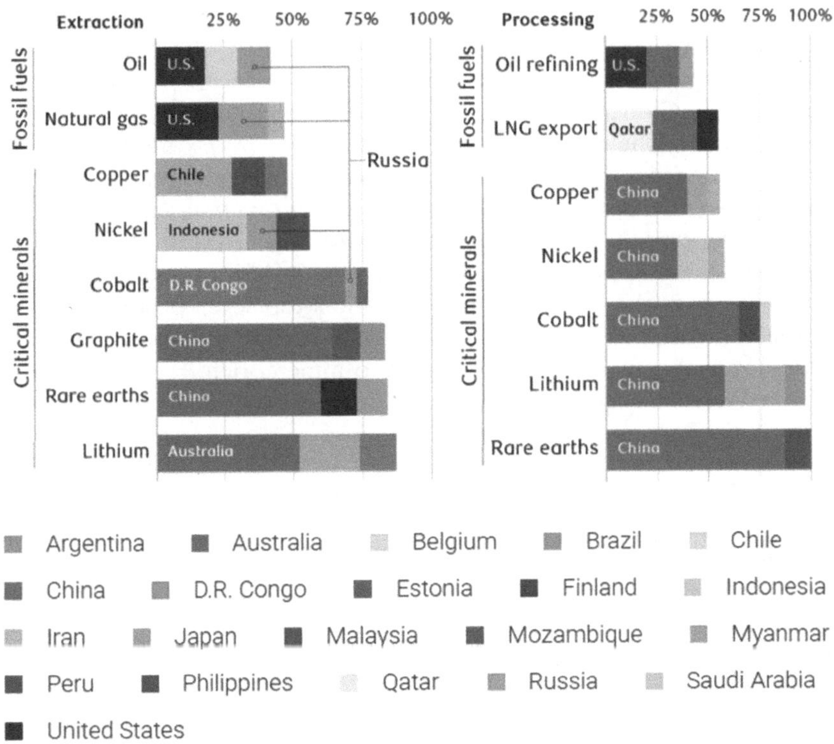

Fig. 4.2 Share of top three producers of selected minerals and fossil fuels, 2019. (Source: RBC Wealth Management based on International Energy Agency (2020), U.S. Geological Survey (2021), World Bureau of Metal Statistics (2020), Adamas Intelligence (2020))

including supply chains for critical minerals and battery components—the so-called friend-shoring.[26]

Similarly, in March 2023, the EU passed the Critical Raw Materials Act (CRMA), which aims to guarantee the EU's "access to a secure and

[26] International Energy Agency, Inflation Reduction Act of 2022 (2025) https://www.iea.org/policies/16156-inflation-reduction-act-of-2022 accessed 23 June 2025; Allan B, Gordon N and Wang C, 'Friendshoring Critical Minerals: What Could the U.S. and Its Partners Produce?' (2023) Carnegie Endowment for International Peace https://carnegieendowment.org/research/2023/05/friendshoring-critical-minerals-what-could-the-us-and-its-partners-produce?lang=en accessed 23 June 2025.

sustainable supply of critical raw materials".[27] The EU Net Zero Industry Act and the Green Deal Industrial Plan complement the CRMA.[28] The EU is also pursuing similar 'friend-shoring' deals with partners ('non-rivals') in the Global South, including Africa, to advance its green transition agenda.[29] The U.K.'s strategy titled, *Resilience for the Future: The United Kingdom's Critical Minerals Strategy*, spells out that it seeks to *"develop our diplomatic, trading and development relationships around the world to improve the resilience of supply to the UK"*. African countries like the Democratic Republic of Congo and South Africa remain central to that strategy, as they rank among the world's top seven producers of 18 critical minerals.[30] Various bilateral trading and development financing assistance packages are being advanced to mineral-rich African countries by the West to secure their strategic interests under Green Minerals Partnership Agreements (Fig. 4.3).

Likewise, emerging players like India, Russia, Turkey and those in the Middle East have also intensified their efforts to secure a share of the global critical mineral supply. For India, the strategic interest in establishing a foothold in the supply chain is driven by rapid domestic economic growth and the desire to avoid heavy reliance on imports, thereby minimising a single point of dependence. India has also set ambitious goals for producing electric vehicles, renewable energy manufacturing and defence. In furtherance of this objective, it is actively working through its state-owned enterprises to seek acquisitions or partnerships with countries such

[27] European Commission, 'Connecting the Democratic Republic of the Congo, Zambia, and Angola to Global Markets through the Lobito Corridor' (2023a) https://international-partnerships.ec.europa.eu/policies/global-gateway/connecting-democratic-republic-congo-zambia-and-angola-global-markets-through-lobito-corridor_en accessed 24 June 2025.

[28] European Commission, 'Critical Raw Materials: ensuring secure and sustainable supply chains for EU's green and digital future' (2023b) https://ec.europa.eu/commission/presscorner/detail/en/ip_23_1661 accessed 23 June 2025.

[29] Vivoda V, 'Friend-shoring and critical minerals: exploring the role of the minerals security partnership' (2023) 100 Energy Research & Social Science 103085; Rizzi A, 'Give and take: How the EU's friend-shoring can improve its relations with the global south' (2023) ECFR https://ecfr.eu/article/give-and-take-how-the-eus-friend-shoring-can-improve-its-relations-with-the-global-south accessed 23 June 2025.

[30] Department for Business, Energy & Industrial Strategy, Resilience for the Future: The United Kingdom's Critical Minerals Strategy (2022) https://assets.publishing.service.gov.uk/media/62f36bafe90e07714288b188/resilience_for_the_future_the_uks_critical_minerals_strategy.pdf accessed 3 May 2025.

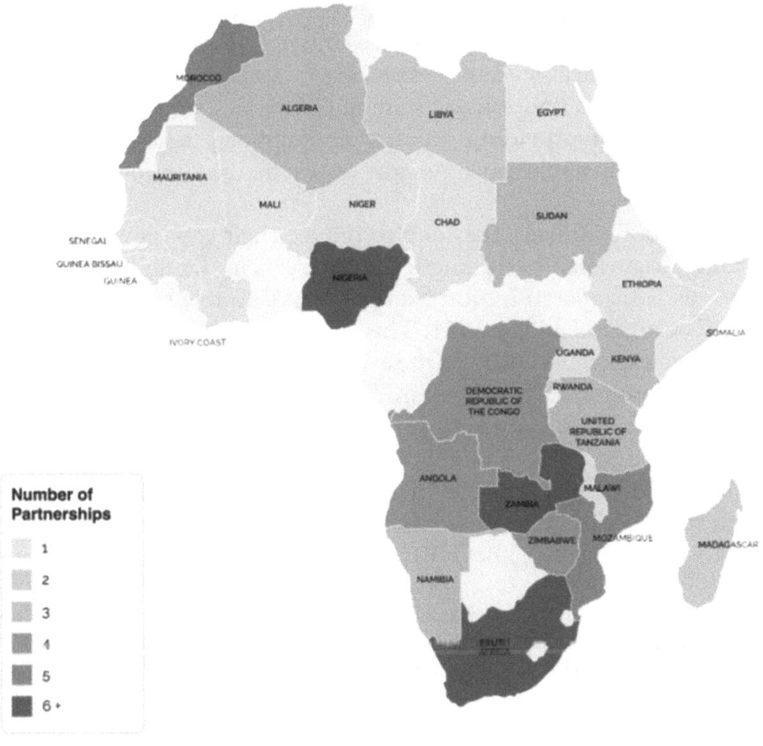

Fig. 4.3 Mapping Africa's green mineral partnerships. (Source: Beuter et al., 2025: p.6)

as the Democratic Republic of the Congo (DRC), Zimbabwe and Namibia. The state enterprise vehicle in India is Khanij Bidesh India Limited (KABIL), a Joint Venture Company among three Government Enterprises: National Aluminium Company Ltd. (NALCO), Hindustan Copper Limited (HCL) and Mineral Exploration & Consultancy Limited (MECL). Securing direct access to critical minerals aligns with its 'Make in India' and energy transition objectives.

Similarly, Turkey's interest in Africa's rich mineral resources aims to enhance its economic and political influence. Turkey's modus operandi involves private Turkish companies working in the construction, energy and logistics sectors. These efforts aim to secure the supply of raw materials for the development of their TOGG electric car project, benefiting the

Turkish local industry.[31] Russia is also expanding its geopolitical influence to counter Western dominance, secure strategic resources, generate revenue and gain political leverage, especially in fragile African states. Russia operates through its state-owned mining companies, coupled with a transactional model of providing security and military support for African countries with unstable governments in exchange for lucrative mining concessions.[32]

For the Gulf countries, they aim to reduce their economic dependence on oil and gas, instead becoming leading players in the future green economy.[33] This is stipulated in, for example, Saudi Arabia's Vision 2030 and the UAE's Centennial Plan 2071. This also includes a target of developing the domestic capacity to process these critical minerals into midstream and downstream products. In this regard, the Gulf Countries are making substantial financial and capital investments in Africa's mining sector, for example, through the signing of bilateral agreements and negotiating direct stakes in copper projects in Zambia.[34] For example, Saudi Arabia, through state-backed ventures, has expressed interest in acquiring mining concessions in Burundi, the DRC and Tanzania.[35] This competition is

[31] Yıldız DT, 'Will the resource potential of critical raw materials used in electric cars in Turkey be sufficient for the domestic automobile factory? – A review' (2025) https://gsm.min-pan.krakow.pl/pdf-197826-124051?filename=Will%20the%20resource.pdf accessed 2 May 2025.

[32] Modern Diplomacy, 'Hired Guns: The Rise of Private Military Industry in Africa' (2025) https://moderndiplomacy.eu/2025/05/03/hired-guns-the-rise-of-private-military-industry-in-africa/ accessed 5 May 2025.

[33] Olawuyi D S, 'Can MENA extractive industries support the global energy transition? Current opportunities and future directions' (2021) The Extractive Industries and Society 8(2) 100685; Al Rawashdeh R and Campbell G, 'Mineral policy in the Gulf Cooperation Council (GCC) countries: the case of Saudi Arabia' (2022) The Extractive Industries and Society 9 101042.

[34] Acheampong T, 'Rules of entry: Cooperative pathways for Europe and Africa on critical minerals' (2024b) https://ecfr.eu/publication/rules-of-entry-cooperative-pathways-for-europe-and-africa-on-critical-minerals/ accessed 2 June 2025; Beuter P, Hofmeyr J, Gabadadze L, Bhuee R and Gnanguênon A, 'Mapping Africa's Green Mineral Partnerships' (2025) APRI https://afripoli.org/mapping-africas-green-mineral-partnerships accessed 23 June 2025; Boafo J, Obodai J, Stemn E and Nkrumah P N, 'The race for critical minerals in Africa: A blessing or another resource curse?' (2024) Resources Policy 93 105046.

[35] Acheampong T, 'Rules of entry: Cooperative pathways for Europe and Africa on critical minerals' (2024b) https://ecfr.eu/publication/rules-of-entry-cooperative-pathways-for-europe-and-africa-on-critical-minerals/; Sarah W, 'The strategies driving the players in competition for Africa's critical minerals' (2024) Atlantic Council https://www.atlanticcouncil.org/blogs/africasource/the-strategies-driving-the-players-in-competition-for-africas-criti-

taking place within a context where African nations are increasingly seeking greater benefits, local value addition and sustainable practices.

4.4 Regional Dynamics: Competition and Cooperation

Intra-African competition and collaboration are key to fully leveraging the potential of the critical minerals sector. At the continental level, the African Continental Free Trade Area (AfCFTA) and the Africa Green Minerals Strategy (AGMS) are two major frameworks that can enhance regional integration of critical mineral value chains. In 2009, the Africa Mining Vision (AMV) was adopted by Heads of State as *"Africa's own response to tackling the paradox of great mineral wealth existing side by side with pervasive poverty"*. This has been translated into the AMV Action Plan, the formation of the African Mineral Development Centre (AMDC), the African Minerals Governance Framework (AMGD), and most recently, the Africa Green Minerals Strategy.[36]

The Africa Green Minerals Strategy (AGMS) is an ambitious strategy founded on four key pillars: advancing mineral development, nurturing people and technical capability, constructing essential value chains and enhancing mineral stewardship and governance (Fig. 4.4). The AGMS has an overarching vision to create "an Africa that harnesses green mineral value-chains for equitable resource-based industrialisation and electrification, creating green technologies and sustainable development to enhance the quality of life of its people".[37]

cal-minerals/; M Procopio & Čok, 'Diversification nations: The Gulf way to engage with Africa' (2025) ECFR https://ecfr.eu/publication/diversification-nations-the-gulf-way-to-engage-with-africa/; Wilson M, 'From the Desert to the Mines: The GCC's Role in Africa's Mineral Development' (2024) https://www.grc.net/documents/66c5a963cce81GCCRoleAfricaMineralDevelopment2.pdf

[36] African Union, 'AMV — Africa Mining Vision' (2009) https://au.int/en/ti/amv/about accessed 23 June 2025; African Union, 'Africa Mining Vision Action Plan' (2011) https://au.int/sites/default/files/documents/29812-doc-action_plan_final_version_jan_2012.pdf; African Union, 'Africa Mining Vision: African Minerals Governance Framework' (2017) https://archive.uneca.org/sites/default/files/PublicationFiles/african_mining_vision_african_mineral_governance_framework.pdf; African Union, 'Africa's Green Minerals Strategy' (2024a) https://au.int/sites/default/files/documents/44539-doc-AGMS_Final_doc.pdf

[37] African Union, 'Africa's Green Minerals Strategy' (2024a) https://au.int/sites/default/files/documents/44539-doc-AGMS_Final_doc.pdf

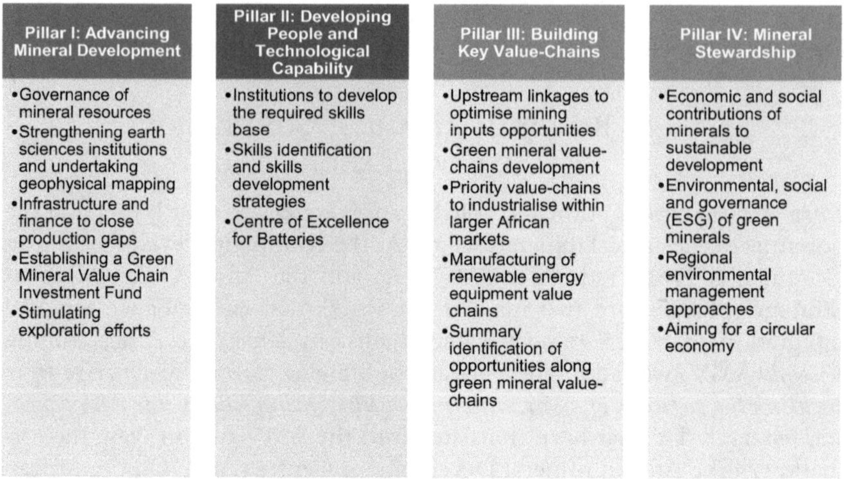

Fig. 4.4 Pillars of the African Green Minerals Strategy. (Source: Authors' construct from African Green Minerals Strategy)

AfCFTA could be a game-changer for regional industrialisation as it creates a single African market encompassing 54 countries by eliminating tariffs and harmonising trade rules. For the critical minerals sector, the AfCFTA could lower the costs of cross-border trade in mineral products and inputs, thereby directly improving project economics and regional value-added initiatives. Investors could source the inputs needed for the battery supply chain from across Africa without punitive costs such as duties. Work is still ongoing to define the rules of origin for complex products, such as those within the CRM's value chain, including mining equipment, battery cells for electric vehicles (EVs) and EVs themselves. Clear rules determining what qualifies as 'Made in Africa' in the context of using African CRMs will enable these products to trade freely within the continent and improve consumer welfare.

Some African countries—such as Zimbabwe, Gabon and Ghana—are pursuing nationalistic strategies to capture more value domestically, by restricting raw exports, imposing higher taxes on concentrate exports and demanding local processing.[38] For example, in 2022, Zimbabwe banned

[38] Briel G, 'Are Critical Minerals Export Bans an Effective Tool for Achieving Diversification in Africa?' (2024) https://www.tralac.org/publications/article/16466-are-critical-miner-

the export of raw lithium ore, and in June 2025, it announced a ban on lithium concentrate exports, effective from 2027, to compel companies to establish domestic refining and battery-grade processing.[39] In January 2020, media reports indicated that a Zambian copper refinery had to suspend operations due to a shortage of cobalt and copper concentrates, resulting from the imposition of a 5% import tax on raw materials from its neighbour, the DRC.[40] While this was a well-intentioned measure to protect domestic value addition by the DRC, it did create friction with Zambia. The ban was subsequently lifted. Such tensions between national sovereignty and the potential economies of scale achievable through regional value chain specialisation remain a core challenge.[41]

Nevertheless, there is growing momentum for more regional cooperation among countries to form joint ventures, share financing and develop integrated value chains, thereby improving economies of scale and scope. This reflects the recognition that while such bans may be noble, each country's insistence on developing its own refinery or manufacturing plant can lead to duplicative and inefficient investment, which makes projects uneconomic, aswhere domestic feedstock is insufficient or processing plants underutilised. Based on the above, there is a growing

als-export-bans-an-effective-tool-for-achieving-diversification-in-africa.html accessed 23 June 2025; TRT Global, 'Gabon latest African country to halt raw export of critical mineral' (2025) https://trt.global/afrika-english/article/afcc644b0e19 accessed 23 June 2025; Africa Center for Strategic Studies, 'Africa's Critical Minerals at a Critical Juncture' (2025) https://africacenter.org/spotlight/africas-critical-minerals-at-a-critical-juncture/; Zhuwakinyu M, 'From geological data to policy, Africa still has big critical-minerals gaps to close' (2025) Mining Weekly https://www.miningweekly.com/article/from-geological-data-to-policy-africa-still-has-big-critical-minerals-gaps-to-close-2025-05-30-1

[39] Reuters, 'Zimbabwe to ban export of lithium concentrates from 2027' (2025) https://www.reuters.com/business/energy/zimbabwe-ban-export-lithium-concentrates-2027-2025-06-10

[40] Reuters, 'ERG suspends Zambia refinery on shortage of cobalt, copper concentrates' (2020) https://www.reuters.com/article/business/finance/erg-suspends-zambia-refinery-on-shortage-of-cobalt-copper-concentrates-idUSKBN1ZM21R; P Barrera, 'ERG's Chambishi Metals Suspends Cobalt Production in Zambia' (2019) Investing News Network https://investingnews.com/daily/resource-investing/battery-metals-investing/cobalt-investing/erg-chambishi-metals-suspends-cobalt-production-zambia/

[41] Karkare P and Medinilla A, Green industrialisation: Leveraging critical raw materials for an African battery value chain (ECDPM Discussion Paper 359, 2023) https://ecdpm.org/application/files/4017/0108/2796/Green-Industrialisation-Leveraging-Critical-Raw-Materials-African-Battery-Value-Chain-ECDPM-Discussion-Paper-359-2023.pdf accessed 23 June 2025.

recognition that regional cooperation on critical raw materials is the way to go if Africa is to break the proverbial 'paradox of plenty'.[42] Regional cooperation can unlock greater value by pooling resources, sharing infrastructure and creating integrated industries. In April 2022, DRC and Zambia signed a Cooperation Agreement to manufacture electric batteries.[43] To actualise the vision of the partnership, they established a DRC-Zambia Battery Council to oversee the implementation of the cooperation agreement. At the same time, the African Export-Import Bank (Afreximbank) indicated its readiness to fund preparations for the project and offer advisory services.

Both countries have established a Transboundary Special Economic Zone (SEZ) for battery manufacturing, which aims to combine their raw materials, expertise and energy resources to produce battery precursors and potentially fully assembled batteries in the near term.[44] Afreximbank and the United Nations Economic Commission for Africa (UNECA) tasked ARISE Integrated Industrial Platforms to undertake a pre-feasibility study on the Transboundary Special Economic Zone.[45] The company submitted the draft pre-feasibility report to the two member States in

[42] Acheampong T, 'Rules of entry: Cooperative pathways for Europe and Africa on critical minerals' (2024b) https://ecfr.eu/publication/rules-of-entry-cooperative-pathways-for-europe-and-africa-on-critical-minerals/ accessed 2 June 2025; Karkare P, 'The EU's partnerships around critical raw materials: Do its ambitions match reality?' (2024) ECDPM https://ecdpm.org/work/eus-partnerships-around-critical-raw-materials-do-its-ambitions-match-reality accessed 23 June 2025; Baranzelli C, Blengini GA, Josa SO and Lavalle C, 'EU--Africa Strategic Corridors and critical raw materials: two-way approach to regional development and security of supply' (2022) 36(9) International Journal of Mining, Reclamation and Environment 607.

[43] UNECA, 'Zambia and DRC Sign Cooperation Agreement to manufacture electric batteries' (2022a) https://www.uneca.org/stories/zambia-and-drc-sign-cooperation-agreement-to-manufacture-electric-batteries accessed 23 June 2025.

[44] UNECA, 'Zambia and DRC to implement an Innovative transboundary battery and Electric vehicle Special Economic Zone' (2024) https://www.uneca.org/stories/zambia-and-drc-to-implement-an-innovative-transboundary-battery-and-electric-vehicle-special accessed 23 June 2025; Olan'g S and Scurfield T, 'The DRC-Zambia Battery Plant: Key Considerations for Governments in 2024' (Natural Resource Governance Institute, 2023) https://resourcegovernance.org/publications/drc-zambia-battery-plant-key-considerations-governments-2024 accessed 23 June 2025.

[45] UNECA, 'Zambia and DRC to implement an Innovative transboundary battery and Electric vehicle Special Economic Zone' (2024) https://www.uneca.org/stories/zambia-and-drc-to-implement-an-innovative-transboundary-battery-and-electric-vehicle-special accessed 23 June 2025.

December 2023. It is expected that the joint SEZ will manufacture cathode precursor materials and lithium-ion batteries for electric vehicles (EVs), leveraging both countries' resources: the DRC's raw material supply (including cobalt, copper and potential hydroelectric power) and Zambia's copper processing infrastructure and industrial expertise.[46]

A materialisation of this DRC-Zambia battery hub is a good example of what regional value chain development could achieve instead of disparate individual country efforts, more so given that no one country on the continent is likely to have all the key factor inputs, such as the resources, energy and skills base, among others. For the international community, such collaboration also sends a signal to investors that cross-border projects could be viable if incentives are aligned and there is strong domestic political and regional institutional support behind them. In January 2023, the United States complemented this effort by signing a memorandum of understanding (MoU) with both the DRC and Zambia to strengthen their battery value chains.[47] An African Centre of Excellence on Electrical Batteries Skills (CAEB) was also opened earlier in April 2022 in Lubumbashi to train technicians in battery and electric car plants. It is affiliated with the University of Lubumbashi and is one of the outcomes of the DRC-Zambia cooperation.

Other regional collaboration initiatives include the Lobito Corridor project, which connects Angola, DRC and Zambia by revitalising the 1300 km brownfield railway from the DRC/Zambian Copperbelt to the Atlantic port of Lobito in Angola.[48] The corridor, when complete, is

[46] Kamwendo E, 'Workshop On "Needs And Financing Mechanisms Of The Special Economic Zone (SEZ) For Battery Electric Vehicles (BEVs) In The Republic Of Zambia"' (2023) https://www.uneca.org/eca-events/sites/default/files/resources/documents/sro-sa/sez-for-bevs-zambia/talking_points_for_bev_sez_workshop_for_zambia_4-5_231004_123757.pdf accessed 23 June 2025.

[47] Burger S, 'US, DRC and Zambia sign MoU to strengthen EV battery value chain' (Mining Weekly, 2023) https://www.miningweekly.com/article/us-drc-and-zambia-sign-mou-to-strengthen-ev-battery-value-chain-2023-01-19 accessed 23 June 2025.

[48] European Commission, 'Connecting the Democratic Republic of the Congo, Zambia, and Angola to Global Markets through the Lobito Corridor' (2023a) https://international-partnerships.ec.europa.eu/policies/global-gateway/connecting-democratic-republic-congo-zambia-and-angola-global-markets-through-lobito-corridor_en accessed 24 June 2025; OECD, 'Background Note -- The Lobito Corridor' (2025) https://www.oecd.org/content/dam/oecd/en/events/2025/04/oecd-emerging-markets-forum/Panel%202_OECD%20EMF%20Background%20Note%20-%20The%20Lobito%20Corridor.pdf accessed 24 June 2025; Karkare P and Byiers B, 'The Lobito Corridor: Between European

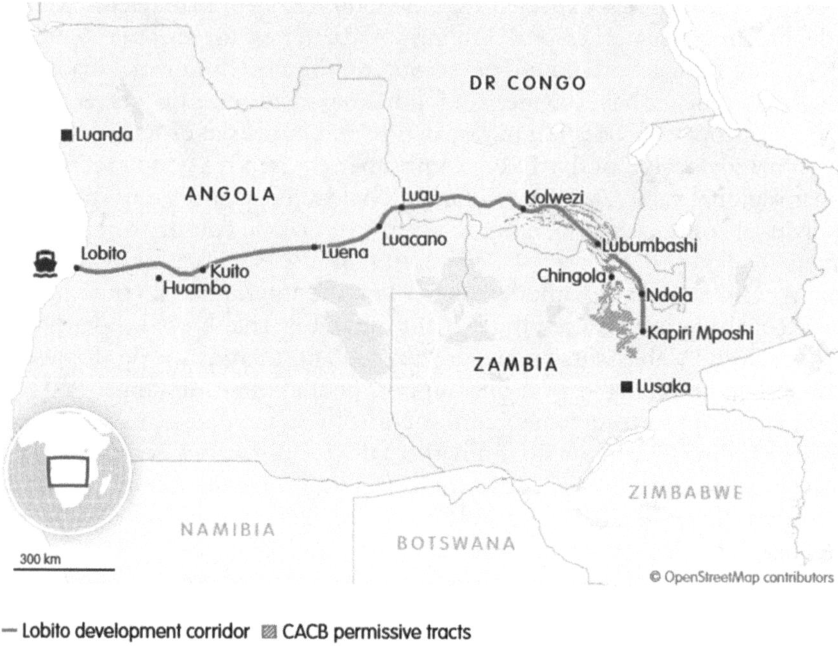

Fig. 4.5 The Lobito Corridor Route. (Source: USGS Global Mineral Resource Assessment, ICGLR, ECFR ecfr.eu)

expected to facilitate the export of copper and cobalt from DRC and Zambia via the shorter westward route (Fig. 4.5). In October 2024, it was reported that KoBold Metals, a U.S. mining company utilising artificial intelligence, had signed a Memorandum of Understanding with the Africa Finance Corporation to become the offtaker that will underpin the commercial viability of the Zambia-Lobito rail project, utilising output from

geopolitics and African agency' (ECDPM Discussion Paper 386, 2025) https://ecdpm.org/application/files/1217/4479/3017/The-Lobito-Corridor-Between-European-geopolitics-African-agency-ECDPM-Discussion-Paper-386-2025.pdf accessed 24 June 2025; Chabala EDW and Hofmeyr J, 'A game changer in flux: recent developments and risks in the Lobito corridor' (April, 2025) https://afripoli.org/a-game-changer-in-flux-recent-developments-and-risks-in-the-lobito-corridor accessed 24 June 2025.

its Mingomba mine.[49] Additionally, the railway line would help move agricultural and other produce to key markets within the country and for exports.

Furthermore, in December 2024, Morocco, the DRC and Zambia launched an electric mobility initiative aimed at creating an Africa-based EV value chain.[50] Morocco, as one of the leading car manufacturers on the continent, with its experience in vehicle assembly, could complement the DRC-Zambia battery SEZ project. For example, the precursors produced in the SEZ could be sent to Morocco for use in making electric vehicles (EVs), which can then be exported to other countries, including Europe, or used domestically on the continent.

4.5 Environmental, Social and Governance (ESG) Considerations

4.5.1 Governance Challenges and Resource Nationalism

Governance challenges and resource nationalism impact the sustainable supply of critical minerals. Large parts of the continent face numerous governance issues, including government stability, transparency, accountability, media freedom and respect for labour and human rights.[51] These governance deficits significantly affect resource governance: utilisation of resource rents for national development, lack of accountability and transparency in mining contracts and rents, corruption, susceptibility to conflict, excessive borrowing, inequality and illicit financial flow.[52] These

[49] US Government, 'FACT SHEET: Partnership for Global Infrastructure and Investment in the Lobito Trans-Africa Corridor' (2024) https://bidenwhitehouse.archives.gov/briefing-room/statements-releases/2024/12/03/fact-sheet-partnership-for-global-infrastructure-and-investment-in-the-lobito-trans-africa-corridor accessed 24 June 2025.

[50] Zouiten S, 'Morocco, DRC, and Zambia Launch Electric Mobility Initiative in Lusaka' (Morocco World News, 2024) <https://www.moroccoworldnews.com/2024/12/167381/morocco-drc-and-zambia-launch-electric-mobility-initiative-in-lusaka/> accessed 24 June 2025; ESI Africa, 'DRC, Zambia and Morocco setting up an e-mobility value chain' (2024) https://www.esi-africa.com/business-and-markets/drc-zambia-and-morocco-setting-up-an-e-mobility-value-chain/ accessed 24 June 2025.

[51] Mo Ibrahim Foundation, Africa's critical minerals (2022) https://mo.ibrahim.foundation/sites/default/files/2022-11/minerals-resource-governance.pdf accessed 2 May 2025.

[52] International Energy Agency, 'Sustainable and Responsible Critical Mineral Supply Chains' (2023) https://www.iea.org/reports/sustainable-and-responsible-critical-mineral-supply-chains/executive-summary accessed 3 May 2025.

Table 4.2 Overview of some CRM-endowed countries in Africa and 2021 resource governance indicators/2024 EITI Scores

Country	*Mineral*	*Brief commentary*
Democratic Republic of Congo	Cobalt, copper	The overall 'poor' score of 36/100 reveals a myriad of resource governance deficits. On the band of a good or bad enabling environment, DRC scored 17/100 in the 2021 rankings
Ghana	Lithium, bauxite	Ghana had a 'satisfactory' score of 69/70 in the overall band. There is still work to be done in revenue management as revealed by a 'weak' score of 54/100. Also, the 'high' 2024 EITI score of 78.5 points also affirms Ghana's progress made over the years
Senegal	Titanium and zirconium	A 'good' score of 75/100 reflects Senegal's overall progress in ensuring a suitable mining and governance environment

Source: Authors' compilation (2025)

factors pose risks to these countries' supply chain of critical minerals. Table 4.2 provides an overview of some African countries with CRMs and their resource governance indicators, as well as EITI Scores.

Rising and escalating political tensions, as well as threats of conflict, are significant concerns for the mining industry and the pursuit of net-zero goals. These security concerns manifest as terrorism, crime, internal conflict and civil unrest. For instance, the DRC, home to substantial reserves of critical minerals, suffers from ongoing political instability, conflict and violence. The eastern regions of the DRC are the most severely affected. There have been reports of deadly attacks on mining operators, their assets and personnel by armed non-state actors.[53] The M23 are often reported as the group that has ignited the resurgence in the conflict in Eastern Congo since late 2021. Their activities are concentrated in the North Kivu province, which contains 70% of the world's coltan reserves and is vital for producing electronic devices, but is also perilous. Furthermore, Mozambique is facing ongoing armed conflict. The Islamic

[53] Mongabay, 'How illicit mining fuels violence in eastern DRC' (2025) https://news.mongabay.com/2025/02/how-illicit-mining-fuels-violence-in-eastern-drc-interview-with-jean-pierre-okenda/ accessed 2 May 2025.

State-Mozambique (IS-M) has established a foothold in the Cabo Delgado province, where most of the country's graphite resources are located.[54]

Resource nationalism also emerges as a significant concern for investors in Africa's mining sector. Resource nationalism refers to a government's desire to assert economic and political control over its natural resources, particularly critical minerals in this context.[55] This concept arises from the perception that foreign governments and investors have an unfair advantage in controlling resource rents. Surging commodity prices, increasing inequality, the rising prominence of critical minerals amid global climate change actions and persistent underdevelopment in these states have all intensified this phenomenon.[56] Governments in some resource-rich countries have employed various strategies to realise their objectives, such as increasing tax shares, raising mining royalties, enhancing local participation, limiting foreign ownership and ensuring compliance with environmental and safety standards. If mismanaged, these actions lead to political, contractual and financial risks for investors. These risks, in turn, undermine the capital and technological investments necessary to establish a sustainable supply chain, while also negatively impacting the socio-economic development of host governments.

4.5.2 *Environmental and Sustainability Concerns*

There are concerns about the adverse impacts of mining projects on the environment, such as water pollution, air pollution, habitat destruction and greenhouse gas emissions.[57] Ensuring the highest regulatory standards is crucial for local community buy-in and strengthening supply chain credentials. The existence of industry standards significantly contributes to addressing various ESG challenges. For instance, the OECD Due Diligence Guidance for Responsible Supply Chains of Minerals focuses on

[54] The Telegraph, 'I watched extremists behead my neighbour: Inside Mozambique's ISIS terror battleground' (2024) https://www.telegraph.co.uk/global-health/terror-and-security/extremists-islamic-state-target-mozambique-oil-gas-reserves/ accessed 4 May 2025.

[55] Poncian J, 'Adopting and practicing resource nationalism in Africa: A case of Tanzania's State Mining Corporation' (2023) Social Sciences & Humanities Open 8(1) 100678.

[56] Ostrowski W, 'The twilight of resource nationalism: From cyclicality to singularity?' (2023) Resources Policy 83103599.

[57] International Energy Agency, 'The Role of Critical Minerals in Clean Energy Transitions' (2021) https://www.iea.org/reports/the-role-of-critical-minerals-in-clean-energy-transitions/reliable-supply-of-minerals accessed 2 May 2025.

Human Rights Due Diligence (HRRD) helps with conflict minerals, supply chain traceability and risk mitigation. It offers a comprehensive framework for companies to recognise and tackle human rights and environmental risks in mineral supply chains, particularly in conflict areas.[58] Additionally, there is the multi-stakeholder Initiative for Responsible Mining Assurance (IRMA) and other standards, which provide a comprehensive and independently verifiable standard for social and environmental performance at mine sites. Governments and investors recognise it as a tool for responsible sourcing. Finally, the International Sustainability Standards Board (ISSB) also provides progressive and comprehensive standards regarding sustainability and financial disclosures.[59]

Alongside national laws and regulations, the UN Guiding Principles on Business and Human Rights (UNGPs) provide a commendable framework for state and non-state actors to implement international best practices through their laws, policies and actions. These Guiding Principles emphasise three core pillars: the State's mandate to protect the fundamental human rights of all, the Corporate Social Responsibility (CSR) to respect human rights and access to effective remedies in the event of breaches.[60] Another key framework from the United Nations is the UN Declaration on the Rights of Indigenous Peoples (UNDRIP), which affirms the collective rights of indigenous peoples, including the right to Free, Prior and Informed Consent (FPIC) for projects affecting their lands and resources. Other international frameworks include the ILO Convention No. 169 (Indigenous and Tribal Peoples Convention), among others.

[58] OECD, 'OECD Due Diligence Guidance for Responsible Supply Chains of Minerals from Conflict-Affected and High-Risk Areas' (2016) https://www.oecd.org/en/publications/2016/04/oecd-due-diligence-guidance-for-responsible-supply-chains-of-minerals-from-conflict-affected-and-high-risk-areas_g1g65996.html accessed 2 June 2025.

[59] ISSB, 'About the International Sustainability Standards Board' (2025) https://www.ifrs.org/groups/international-sustainability-standards-board/ accessed 24 June 2025.

[60] United Nations, 'Guiding Principles on Business and Human Rights' (2011) https://www.ohchr.org/sites/default/files/documents/publications/guidingprinciplesbusinesshr_en.pdf accessed 2 June 2025.

4.6 Conclusion

As global demand for critical minerals soars, African governments have recognised the need to assert their interests strategically. There is a growing consensus that Africa should move beyond its current role as a mere exporter of raw minerals by implementing policies and strategies aimed at local downstream value addition, supported by a strong political commitment to sound economic policies that promote sustainable development and ensure that the vast wealth of critical minerals benefits its citizens.

In this light, steps are being taken on the continental front to secure the mineral wealth dividend for its people. Spearheading these efforts is the African Union, which developed "The Africa Mining Vision" and the Africa Green Minerals Strategy. On the national front, some countries have introduced policies to advance their interests. For instance, the DRC, Namibia and Zimbabwe have implemented restrictions or, in some cases, an outright ban on the export of certain raw critical minerals, advocating for local processing. Additionally, Ghana, a new entrant in the lithium sector, has enacted a policy prohibiting the export of lithium extracted in the country in its raw state.

Other powerful geopolitical actors are also engaging with African countries through numerous bilateral agreements. Western nations are determined to counter China's dominance in the midstream and downstream segments of the critical minerals supply chain by developing alternatives, including by expanding sourcing and investment in Africa. This competition will likely manifest through strategic investments, partnerships, diplomatic and economic pressure, and an emphasis on technological leadership. For example, initiatives like the US-led Minerals Security Partnership (MSP) aim to diversify supply chains away from China. Nevertheless, new geopolitical poles and alliances could also emerge. Like-minded, mineral-rich nations might form 'mineral blocs' and partnerships to assert their power through their collective mineral strength. Market volatility and price fluctuations are expected to intensify.

Public-private partnerships (PPPS) for financing African mining projects have gained traction as a sustainable approach to addressing infrastructure challenges. PPPs are a viable financing mechanism because they combine resources, expertise, institutional knowledge and networks of public institutions with the innovation, dynamism and efficiency of private enterprises. There are several groundbreaking and prominent examples of such partnerships. For instance, the Lobito Corridor, which connects the trio of Angola, the Democratic Republic of the Congo (DRC) and Zambia to other trade markets through the port of Lobito.

CHAPTER 5

Ensuring Sustainability in Mining for the Energy Transition

Abstract This chapter examines sustainability and sustainable development in the mining sector, with particular focus on critical minerals essential for the global energy transition. It distinguishes sustainability as a long-term resilience framework and sustainable development as the policy-driven pathway to balance economic growth, environmental protection and social equity. Anchored in the UN Sustainable Development Goals, especially SDGs 7 and 13, the chapter highlights the mineral dependence of low-carbon technologies. It explores disparities between developed and developing countries in implementing sustainability standards, emphasising governance and capacity challenges in resource-rich regions. The chapter analyses sustainability in mining through environmental, social and economic lenses, underscoring the growing role of responsible mining standards and ESG frameworks. It evaluates voluntary and semi-mandatory standards, their governance, credibility and harmonisation efforts such as the Consolidated Mining Standard Initiative. The legal architecture governing mining sustainability is reviewed, revealing fragmented international regulation and stronger reliance on national legal frameworks. Particular attention is given to African constitutional and legislative approaches to mineral governance. The chapter critically assesses the Social Licence to Operate as a determinant of project legitimacy and community acceptance. It further examines the role of artisanal and

© The Author(s), under exclusive license to Springer Nature Switzerland AG 2026

R. J. Heffron et al., *Revolution in the Global Critical Minerals Sector*, Just Transitions,
https://doi.org/10.1007/978-3-032-26366-7_5

small-scale mining, highlighting its socio-economic importance alongside governance, health and environmental challenges. Environmental degradation, pollution and occupational health risks linked to critical mineral extraction are analysed in depth. Recent mining law reforms in Africa are reviewed through comparative case studies. The chapter concludes that sustainable mining requires robust regulation, credible standards, inclusive governance and international cooperation. Ultimately, it argues that responsible mining is central to achieving a just, resilient and low-carbon global economy.

Keywords Global critical minerals treaty • International mineral governance • Critical minerals policy • Global cooperation • United Nations policy • Just transition • Climate-aligned mining • International law and minerals • Economic development • Civil society engagement • Mining sector reform • Sustainable resource governance • Justice in mineral development

5.1 Introduction: Sustainable Development

Sustainability and sustainable development are interrelated concepts that aim to balance the fulfilment of current human needs with the preservation of resources and ecosystems for future generations. At its foundation, sustainability emphasises the long-term resilience and viability of environmental, social and economic systems. Sustainable development, meanwhile, focuses on the specific strategies, policies and practices needed to achieve this balance, ensuring that economic growth, environmental protection and social equity are pursued simultaneously and harmoniously.

These concepts are embodied in the United Nations Sustainable Development Goals (SDGs), a universal blueprint adopted in 2015 to guide global efforts towards sustainability by 2030.[1] Several SDGs are particularly relevant to natural resources and the global energy transition, though others are equally essential. Regarding the energy sector, the UN SDG 7 advocates for universal access to reliable, modern and sustainable energy for all. Achieving this goal requires rapid expansion of renewable energy infrastructure, which is heavily dependent on critical minerals such as lithium, cobalt, nickel and rare earth elements. The UN SDG 13, on

[1] United Nations, 'The Sustainable Development Goals' https://sdgs.un.org/goals accessed 23 January 2025.

the other hand, is focused on climate action and urges countries to take urgent steps to combat climate change and its impacts.[2] Reducing greenhouse gas emissions and transitioning to low-carbon economies requires the deployment of green technologies, including electric vehicles, solar panels, and wind turbines, which rely on a stable, sustainable supply of minerals.

While the goals are universal, the implementation of sustainability standards and practices varies significantly between developed and developing countries. Developed nations often have more robust regulatory frameworks, financial resources, and technological capabilities to enforce environmental and social standards. In contrast, developing countries, many of which are rich in mineral resources, may face challenges such as weak governance, limited infrastructure and socio-economic pressures that complicate the enforcement of sustainable mining and development practices.[3]

This disparity can lead to unsustainable resource extraction in regions least equipped to manage their environmental and social consequences, even though these regions are essential to the global supply chains for green technologies. As such, achieving global sustainability will require not only technological and policy innovation but also international cooperation, investment and capacity-building to ensure that all countries, regardless of development status, can contribute to and benefit from a more sustainable future. While there are various sustainability issues in mining, this chapter will focus on environmental and social issues, with particular attention to the legal and regulatory environment.

This chapter examines the vital role of sustainability in mining, especially in advancing the global energy transition. It distinguishes sustainability and sustainable development as interconnected concepts focused on meeting present needs while preserving resources and ecosystems for the future. Emphasising resilience across environmental, social and economic systems, sustainable development entails policies that balance economic growth with ecological protection and social equity. The chapter begins by defining key sustainability concepts relevant to mining and

[2] United Nations, 'The Sustainable Development Goals' https://sdgs.un.org/goals accessed 23 January 2025.

[3] Nalule VR. Regulation of mining in Africa. Mining and the Law in Africa: Exploring the social and environmental impacts. 2020:19–50.

energy, then reviews legal and regulatory frameworks, contrasts practices in developed and developing countries and addresses governance challenges in resource-rich but underdeveloped regions.

5.2 Sustainability in Mining for the Energy Transition: The Role of Mining Standards

Sustainability in mining refers to the pursuit of long-term economic, environmental and social balance in the extractive industry. As the global demand for minerals surges, especially those critical to renewable energy technologies and digital infrastructure, the mining sector is under increasing pressure to operate responsibly. Sustainability in this context means more than reducing environmental harm; it requires companies to contribute meaningfully to local communities, respect human rights and promote lasting economic development.

The transition to a low-carbon economy hinges on the availability of key minerals like lithium, cobalt, copper and rare earth elements. These minerals are indispensable to achieving global climate goals, particularly Sustainable Development Goals (SDGs) such as SDG 7 on affordable and clean energy and SDG 13 on climate action.[4] Yet, the process of extracting these resources often carries significant environmental and social risks, making sustainable mining practices a necessity rather than an option.

Efforts to embed sustainability into mining operations are evident in a range of practices. These include the reuse of tailings, extracting remaining metals from waste materials to minimise environmental impact, as well as improved waste and water management systems that prevent pollution and reduce operational risks. Mining companies are also increasingly engaging with local communities, ensuring that they benefit from resource development through employment, infrastructure and shared decision-making.

To guide and measure these efforts, a growing number of responsible mining standards and sustainability frameworks have emerged. These standards provide structured approaches for companies to manage environmental, social and governance (ESG) risks and align their operations with international best practices. They address key areas such as emissions

[4] United Nations, 'The Sustainable Development Goals' https://sdgs.un.org/goals accessed 23 January 2025.

reduction, biodiversity protection, worker health and safety, ethical conduct, transparency and social inclusion. Some standards are voluntary, while others are semi-mandatory, depending on investor or stakeholder requirements. Regardless of their legal standing, they serve as essential tools to encourage improved performance, build public trust and support the broader sustainability agenda.

Mining standards vary widely in type and focus. Supply chain due diligence standards ensure minerals are sourced ethically, avoiding conflict and human rights abuses. Performance or compliance-based standards set measurable benchmarks at mine sites, often verified by third-party audits. Principle-based standards provide broader guidance on ethical and sustainable behaviour, aligned with global norms like the UN Guiding Principles on Business and Human Rights.

The credibility of a mining standard depends on its governance, assurance mechanisms and transparency. Multi-stakeholder development processes lend greater legitimacy by incorporating views from governments, companies, civil society and local communities. Independent third-party certification enhances trust by providing objective assessments, while public disclosure and value chain traceability are essential for accountability and stakeholder confidence.

Mining companies adopt these standards to meet growing market and investor demands for responsible sourcing, gain strategic advantages, mitigate risks, and improve practices. Standards also help align operations with global investment criteria, such as those of the International Finance Corporation (IFC).

However, implementation is uneven globally. Developed countries typically have stronger regulations and capacity, while many mineral-rich developing countries face governance, infrastructure, and stability challenges. This disparity can weaken sustainability efforts and increase risks of environmental harm and social unrest, despite these regions' growing role in global supply chains.

The proliferation of standards has created complexity, overwhelming especially smaller companies and causing reporting fatigue. In response, initiatives like the Consolidated Mining Standard Initiative (CMSI), launched in 2024, seek to harmonise key standards, combining frameworks such as the International Council on Mining and Metals (ICMM), The Copper Mark, Towards Sustainable Mining (TSM) and the World Gold Council into a unified, streamlined system. This aims to reduce duplication, simplify compliance and encourage wider adoption across the industry.

Other notable initiatives include the Initiative for Responsible Mining Assurance Standard 2.0, which broadens environmental and social criteria across all mining stages from exploration to closure, with enhanced focus on stakeholder engagement, transparency and auditability; the United Nations High-Level Panel on Critical Energy Transition Minerals, which has set guiding principles for responsible value chains emphasising human rights, environmental protection and governance; and the Mining Policy Framework by the Intergovernmental Forum on Mining, Minerals, Metals and Sustainable Development, which aids governments in developing regulatory frameworks aligned with sustainability goals.

Governments play a pivotal role in shaping the sustainability landscape. While voluntary standards are valuable, they are not substitutes for robust regulatory systems. Governments are encouraged to participate in the development of standards to ensure national priorities are reflected. They can also use voluntary standards to strengthen or supplement their own regulations, especially in contexts where regulatory capacity is limited. Promoting data sharing between standard-setting bodies and regulators enhances oversight and helps align public policy with industry practices. Furthermore, by anticipating emerging issues flagged by voluntary standards, governments can future-proof their regulatory frameworks and position themselves as attractive destinations for responsible investment.

As sustainability expectations rise and the industry evolves, it is clear that responsible mining standards will remain central to ensuring that the minerals required for the energy transition are extracted ethically and sustainably. Yet, to achieve real impact, these standards must be credible, globally coherent, and implemented equitably across diverse operating contexts. With stronger coordination among industry actors, policymakers and communities, the mining sector can play a transformative role in advancing a more just, inclusive and sustainable global economy.[5]

[5] International Institute for Sustainable Development, 'Navigating Global Sustainability Standards in the Mining Sector' (Intergovernmental Forum on Mining, Minerals, Metals and Sustainable Development, November 2024) https://www.iisd.org/system/files/2024-11/global-sustainability-standards-mining.pdf accessed 1 March 2025.

5.3 Legal Architecture for Mining Sustainability

5.3.1 Introduction

Mining is vital to global development, supplying raw materials for industries including clean energy. However, traditional practices have caused environmental harm, social disruption and health risks. In response, sustainable mining is gaining momentum, focusing on minimising impacts, promoting social responsibility and ensuring long-term viability.

Many operations now use renewable energy, advanced water management and waste-reduction techniques such as tailings reprocessing and land reclamation. Biodiversity plans and circular economy principles are increasingly adopted. Technological advances such as electric vehicles, automation and remote sensing help reduce emissions and improve safety, while Environmental Impact Assessments (EIAs) identify and mitigate risks early.

Ethical sourcing and strong environmental, social and governance (ESG) standards are shaping responsible supply chains. Social sustainability is equally important, with community engagement, fair compensation and inclusive development key to securing a Social Licence to Operate (SLO). These efforts are essential not only for ethics but also for long-term project success. At this juncture, it is worth examining the practicality of the social licence to operate.

5.3.2 Social Licence to Operate

Although there is no universal definition of what a social licence to operate (SLO) entails, it can be considered to exist when a mining project is seen as having the ongoing approval and broad acceptance of society to conduct its activities.[6] While it is unwritten, in practice, it has governed the relationship between corporations and host communities. It plays a major role in determining which corporate activities societies will accept or reject, whether or not these are embodied in the law.[7] Consequently, it is

[6] J Prno and DS Slocombe, 'Exploring the Origins of "Social License to Operate" in the Mining Sector: Perspectives from Governance and Sustainability Theories' (2012) 37 *Resources Policy* 346.

[7] I Thomson and RG Boutilier, 'Social License to Operate' in Peter Darling (ed), *SME Mining Engineering Handbook* (3rd edn, Society for Mining, Metallurgy, and Exploration 2011) vol 1, 1779–96.

imperative to note that SLO is distinct from the legal licence. As such, society can withhold its SLO even though the company has already obtained the legal licence from the government.[8] A question may arise as to what constitutes a 'society'. It is worth noting that whereas the general public, government and community can constitute a society, it is often local communities that serve as key arbiters in the SLO process, given their proximity to projects, sensitivity to effects and ability to affect project outcomes.[9]

Although SLO is an emerging concept and might appear theoretical, in practice, we have witnessed its impact through protests, blockades and the retraction of government-issued mining permits following clashes between mining companies and local communities. The need for SLO has become increasingly crucial in recent years, driven by the growing demand for critical minerals for the energy transition. Reliable data show that mining these resources has increased social unrest in countries such as Indonesia, Peru and Serbia, among others, primarily due to the social and environmental harm associated with the exploitation of critical minerals.[10] For instance, in Indonesia, nickel mining was linked to human rights abuses.[11] According to reports, it is estimated that between 2021 and 2023 alone, there were 334 incidents of violence or protest linked to mining copper, cobalt, lithium and nickel in the top 10 producing countries.[12] As demand and production of transition minerals are expected to increase, related social protests are also likely to rise. For instance, estimates show that, between 2021 and 2028, copper mining is forecast to increase by over 25%, cobalt by over 100%, lithium by over 300% and nickel by over 75%

[8] K Komnitsas, 'Social License to Operate in Mining: Present Views and Future Trends' (2020) 9 *Resources* 79.

[9] J Prno and DS Slocombe, 'Exploring the Origins of "Social License to Operate" in the Mining Sector: Perspectives from Governance and Sustainability Theories' (2012) 37 *Resources Policy* 346.

[10] Climate Rights International, 'Indonesia: Nickel Industry Harming Human Rights and the Environment' (5 June 2025) https://cri.org/indonesia-nickel-industry-harming-human-rights-and-the-environment/ accessed 2 June 2025.

[11] Climate Rights International, 'Indonesia: Nickel Industry Harming Human Rights and the Environment' (5 June 2025) https://cri.org/indonesia-nickel-industry-harming-human-rights-and-the-environment/ accessed 2 June 2025.

[12] Global Witness, 'Critical Mineral Mines Tied to 111 Violent Incidents and Protests on Average a Year' https://globalwitness.org/en/campaigns/transition-minerals/critical-mineral-mines-tied-to-111-violent-incidents-and-protests-on-average-a-year/ accessed 16 June 2025.

globally.[13] This projected increase in production will most likely escalate the environmental and social impacts, especially in emerging and developing resource-rich producing countries, which, according to recent data, experienced nearly 90% of protest and violence events related to critical minerals.[14]

Reflecting on the above, it is essential to ensure that local communities are well informed about the different ways they can leverage SLO to make mining critical minerals a blessing rather than a curse in their communities. In addition to the legal licence, it becomes imperative for mineral developers to obtain the SLO to avoid potentially costly conflicts and exposure to social risks. Scholars have identified three normative components of the SLO, including legitimacy, credibility and trust.[15] Additionally, there are four levels of SLO, including withdrawal, acceptance, approval and identification with the project psychologically.[16] They argue that the shift from legitimacy to credibility and ultimately, to trust, depends on building social capital through the company's relationships with local stakeholders. A community's ability to grant a meaningful Social Licence to Operate (SLO) relies on its network's social capital, which stems from three sources: (1) structural (the company's integration into the community network), (2) relational (reciprocity, shared identity, and trust) (3) and cognitive (mutual understanding and agreement).[17]

A question may arise as to whether the SLO for large-scale mining differs from the SLO for artisanal and small-scale mining. For LSM,

[13] Global Witness, 'Critical Mineral Mines Tied to 111 Violent Incidents and Protests on Average a Year' https://globalwitness.org/en/campaigns/transition-minerals/critical-mineral-mines-tied-to-111-violent-incidents-and-protests-on-average-a-year/ accessed 16 June 2025.

[14] Global Witness, 'Critical Mineral Mines Tied to 111 Violent Incidents and Protests on Average a Year' https://globalwitness.org/en/campaigns/transition-minerals/critical-mineral-mines-tied-to-111-violent-incidents-and-protests-on-average-a-year/ accessed 16 June 2025.

[15] Thomson I and Boutilier RG, 'Social License to Operate' in Peter Darling (ed), *SME Mining Engineering Handbook* (3rd edn, Society for Mining, Metallurgy, and Exploration 2011) vol 1, 1779–96.

[16] Thomson I and Boutilier RG, 'Social License to Operate' in Peter Darling (ed), *SME Mining Engineering Handbook* (3rd edn, Society for Mining, Metallurgy, and Exploration 2011) vol 1, 1779–96.

[17] Thomson I and Boutilier RG, 'Social License to Operate' in Peter Darling (ed), *SME Mining Engineering Handbook* (3rd edn, Society for Mining, Metallurgy, and Exploration 2011) vol 1, 1779–96.

communities deal directly with mining corporations, and their demands and expectations vary significantly from those of ASM. This is because, in most cases, the same local communities are involved in ASM. Although the literature linking SLO to ASM is limited, this topic is worth exploring, especially given that most ASM operations also involve critical minerals. In the subsection below, we will briefly explore SLO in relation to ASM.

5.3.3 *Artisanal and Small-Scale Mining (ASM) and the Social Licence to Operate in Africa*

Artisanal and Small-Scale Mining (ASM) has become increasingly prominent in Africa as more people rely on it for their livelihoods. ASM typically involves individuals or small groups mining independently using basic, often manual, tools.[18] While artisanal mining is usually subsistence-based, small-scale mining may involve hired labour and more organised operations.[19] Despite its informal nature, ASM plays a significant economic and social role in many African countries, employing hundreds of thousands and supporting millions more indirectly.[20]

However, ASM also presents governance challenges that affect its ability to gain a Social Licence to Operate (SLO). In many African nations, the sector operates outside formal legal frameworks, leading to environmental degradation, health risks and social tensions. Governments have responded with varying degrees of intervention, including temporary bans, as seen in Uganda, the Democratic Republic of Congo (DRC) and Mali, often aimed at restructuring or formalising the sector.[21] While these measures are intended to address regulatory gaps, they can undermine trust and disrupt livelihoods when not paired with inclusive consultation or support mechanisms.

Securing SLO for ASM requires addressing its broader social and environmental impacts. Issues such as lack of reliable data, gender disparities

[18] Hilson G, McQuilken J. Four decades of support for artisanal and small-scale mining in sub-Saharan Africa: a critical review. The Extractive Industries and Society. 2014 Mar 1;1(1):104–18.

[19] Hentschel T, Hruschka F, Priester M. Global report on artisanal and small-scale mining. Mining, minerals and sustainable development. 2002 Jan;70(1).

[20] Nalule VR. Mining and the Law in Africa: Exploring the social and environmental impacts. Springer Nature; 2019 Nov 22.

[21] Nalule VR. Mining and the Law in Africa: Exploring the social and environmental impacts. Springer Nature; 2019 Nov 22.

in earnings and the use of unsafe, rudimentary mining methods contribute to community and stakeholder concerns. At the same time, ASM contributes substantially to national mineral output and local economies—as illustrated by artisanal gold mining accounting for around 12% of global gold production and 10% of Mali's gold output.[22]

To build and maintain an SLO, efforts must focus on formalising ASM in ways that protect community interests, improve working conditions, ensure environmental safeguards and create transparent governance systems. Recognising ASM's socio-economic significance while addressing its challenges is key to ensuring it operates with legitimacy, credibility and long-term community trust. Moreover, there is now an increase in the number of ASM miners involved in critical minerals. It is important to note that the informality of artisanal and small-scale mining (ASM) poses both challenges and opportunities for critical minerals. Many ASM operations remain unlicensed due to costly or complex permitting requirements, limited geological data, lack of capital and poor access to equipment. Existing regulations often fail to meet ASM's needs, hindering formalisation efforts. However, greater recognition of ASM's significant role, such as Indonesia's 2645 illegal ASM sites across 30 provinces,[23] the largely informal artisanal cobalt mining in the Democratic Republic of Congo overlapping large-scale concessions and Peru's growing artisanal copper mining in the Apurimac region, could drive reforms.[24] Formalising ASM in these key areas is essential to integrating it into sustainable critical mineral supply chains.

In conclusion, the Social Licence to Operate (SLO) for critical minerals is dynamic, intangible and shaped by local social, economic and environmental conditions. As demand for these minerals grows, especially for clean energy transitions, community expectations and scrutiny intensify. While agreements like Impact and Benefit Agreements offer some stability, they do not guarantee lasting acceptance, which remains subject to change. The SLO exists within a broader social–ecological system, where evolving

[22] Nalule VR, *Introduction to Mining in Africa*, in *Mining and the Law in Africa: Exploring the Social and Environmental Impacts* (2020) 1–7.

[23] IISD, *Artisanal and Small-Scale Mining of Critical Minerals* (2024) https://www.iisd.org/publications/report/artisanal-small-scale-mining-critical-minerals accessed 18 June 2025.

[24] IISD, *Artisanal and Small-Scale Mining of Critical Minerals* (2024) https://www.iisd.org/publications/report/artisanal-small-scale-mining-critical-minerals accessed 18 June 2025.

stakeholder relationships and contextual factors directly influence project outcomes.

5.3.4 *Health and Environmental Challenges in Critical Mineral Extraction*

Basically, mining operations significantly impact the environment through high water consumption, land degradation, pollution and greenhouse gas emissions. Water is used extensively in mineral processing, for dust suppression, separation and waste management, often reducing freshwater availability for local communities and leading to contamination when toxic materials from tailings, such as mercury, cyanide or arsenic, enter water sources.[25]

Poorly managed artisanal and small-scale mining (ASM), common in developing countries, further exacerbates pollution due to inadequate safety and environmental controls.[26] Another issue is land degradation, due to excavation, deforestation and infrastructure development such as roads and camps, disrupting ecosystems, especially in sensitive regions like tropical rainforests. Additionally, mining contributes to greenhouse gas emissions through loss of vegetation and the use of fossil fuels, with some materials, like diamonds, having a far higher carbon footprint per kilogram than more common minerals like iron.[27] Despite international regulations, environmental harm from mining remains a pressing issue, particularly where enforcement is weak.

The extraction of critical minerals, especially through artisanal and small-scale mining (ASM), poses significant health and environmental risks. In many ASM communities, miners operate without essential protective equipment such as gloves, boots and goggles. The use of rudimentary tools and methods, combined with poor waste management and the

[25] Sammy Witchalls, 'The Environmental Problems Caused by Mining' (Earth.org, 3 April 2022) https://earth.org/environmental-problems-caused-by-mining/ accessed 23 June 2025

[26] Earth.org, 'Environmental Problems Caused by Mining' https://earth.org/environmental-problems-caused-by-mining/ accessed 23 June 2025.

[27] Earth.org, 'Environmental Problems Caused by Mining' https://earth.org/environmental-problems-caused-by-mining/ accessed 23 June 2025

handling of hazardous chemicals like mercury, results in serious health hazards and environmental degradation.[28]

For example, in gold ASM operations, mercury is commonly used for amalgamation, a process that releases toxic substances into the environment and endangers miners' health. During a field visit to a gold mine in Mubende, Uganda, miners were observed using mercury without protective gear, even though they were aware of its harmful effects and expressed a willingness to adopt safer practices. Similar conditions have been reported in Ghana and Mali.

Beyond chemical exposure, the unsafe nature of ASM methods often leads to fatal accidents. In February 2019, more than ten artisanal miners drowned in Zimbabwe after heavy rains flooded inactive shafts. In January 2024, over 70 miners were reported dead in Mali due to an artisanal mine collapse.[29] These incidents highlight the urgent need for improved safety infrastructure and enforcement of mining regulations. In response to increasing accidents and environmental risks, Mali suspended artisanal gold mining during the 2025 June–September rainy season to prevent landslides.[30] The decision followed meteorological forecasts of heavy rainfall and a prior move in March to halt new ASM permits for foreigners after fatal incidents.[31] Other African countries also adopted similar safety-driven measures.

Large-scale mining (LSM) is not exempt from these challenges. Despite perceptions of higher safety standards, some LSM operations also fail to provide adequate protective gear, with weak enforcement of safety protocols. While health and safety standards are embedded in mining legislation, the key issue remains enforcement.

[28] Nalule VR. Social and environmental impacts of mining. Mining and the Law in Africa: Exploring the social and environmental impacts. 2020:51–81.

[29] Al Jazeera, 'More than 70 Dead in Artisanal Mine Collapse in Mali' (24 January 2024) https://www.aljazeera.com/news/2024/1/24/more-than-70-dead-in-artisanal-mine-collapse-in-mali accessed 2 April 2025

[30] Africanews, 'Mali Suspends Artisanal Gold Mining Permits for Foreign Companies After Series of Accidents' (7 March 2025) https://www.africanews.com/2025/03/07/mali-suspends-artisanal-gold-mining-permits-for-foreign-companies-after-series-of-accident// accessed 2 April 2025.

[31] Africanews, 'Mali Suspends Artisanal Gold Mining Permits for Foreign Companies After Series of Accidents' (7 March 2025) https://www.africanews.com/2025/03/07/mali-suspends-artisanal-gold-mining-permits-for-foreign-companies-after-series-of-accident// accessed 2 April 2025.

Addressing these challenges requires sustainable solutions, including access to modern energy and technologies that enable safer, more efficient mineral extraction. Strengthening regulatory frameworks and supporting miners with education, infrastructure and protective equipment are essential to aligning the critical minerals sector with sustainability goals.

5.3.5 *Further Ongoing International Legal Issues in the Global Mining Sector*

The legal governance of mining remains fragmented, especially at the international level. The 1972 Stockholm Declaration, which established the principle of national sovereignty over natural resources, has guided international law for decades. This principle was reaffirmed in major United Nations summits, including the 1992 Earth Summit, the 2002 Johannesburg Summit and the 2012 Rio+20 Conference. While these events recognised the role of mining in development and called for stronger environmental and social safeguards, they failed to produce a binding international treaty dedicated to mining. Instead, mining is regulated through a patchwork of international investment treaties, human rights instruments, environmental conventions and national laws, often with inconsistent implementation and limited protections for local communities and ecosystems. This fragmented framework poses significant challenges for the responsible sourcing of critical minerals such as lithium, cobalt and rare earth elements, which are essential for the clean energy transition. In addition, national laws and regulations seem more effective than regional and international laws.

At the national level, the starting point is the country's constitution. The Constitution, being the supreme law of the land, should set the pace regarding the regulation and governance of the mining sector. For instance, Uganda's 1995 Constitution (as amended) under Article 244 emphasises that mineral exploitation should benefit the people, considering the interests of landowners, local governments and the state.[32] The Constitution establishes state sovereignty over natural resources, with the state acting as a custodian on behalf of the people.

[32] The 1995 Ugandan Constitution (as amended).

In Zambia, the Constitution outlines the principles regarding the ownership and regulation of natural resources, including minerals, within its territory. Article 59(1) of the Zambian Constitution states, "*All natural resources, including mineral resources, within the Republic, are vested in the President who holds them in trust for the people of Zambia*".[33] This provision clearly establishes that mineral resources are owned by the state and controlled by the President on behalf of the people.

Additionally, Article 60 of the Zambian Constitution asserts, "*The State shall manage the country's natural resources to ensure sustainable development and benefit the people*". This provision reinforces the state's duty to ensure that the exploitation of natural resources, including minerals, is carried out in a manner that supports the country's socio-economic development while safeguarding long-term sustainability.

In Ghana, the Constitution clearly outlines the state's role in the ownership and management of natural resources, including minerals. The 1992 Constitution of Ghana establishes the principle of state ownership over the country's natural resources, including minerals, while also ensuring that these resources are used for the benefit of the people.

- Article 257(1) of the Ghanaian Constitution states: "*All minerals in their natural state in, under, or upon any land in Ghana, shall be the property of the Republic, and shall be vested in the President on behalf of and in trust for the people of Ghana*". This provision affirms that the government, represented by the President, holds ownership of mineral resources in trust for the people of Ghana.[34]

In South Africa, the Minerals and Petroleum Resources Development Act (MPRDA) of 2002, which aligns with the constitutional framework, specifies the state's custodianship over mineral resources. According to Section 3(1) of the MPRDA, "*All minerals and petroleum in the Republic are the common heritage of all the people of South Africa and the state is the custodian thereof for the benefit of all South Africans*".[35] This provision establishes that the state holds mineral resources in trust for the people.

In Kenya, the 2010 Constitution also affirms the state's role in managing mineral resources for the benefit of its citizens. Article 62(2)

[33] The 1991 Zambian Constitution (as amended).

[34] The Constitution of the Republic of Ghana, 1992.

[35] The Minerals and Petroleum Resources Development Act (MPRDA) of 2002, South Africa.

provides that "*Mineral resources and mineral oils are vested in the national government, and their management and exploitation shall be regulated by the national government*".[36] This provision places all mineral resources under state ownership, empowering the government to regulate their extraction, exploitation and distribution for national benefit.

Additionally, relevant laws, such as environmental, taxation, land and employment laws, play key roles in the governance of the sector. However, governance gaps exist at the national, regional and international levels.

To address governance gaps, voluntary standards like the Initiative for Responsible Mining Assurance (IRMA), Towards Sustainable Mining (TSM), the Copper Mark and the Responsible Gold Mining Principles promote strong environmental, social and governance (ESG) practices. The Intergovernmental Forum on Mining, Minerals, Metals and Sustainable Development (IGF) supports over 75 countries in strengthening mining policy frameworks.[37]

Meeting rising demand for critical minerals sustainably will require more than operational improvements; it demands robust national regulations, coordinated international frameworks and inclusive governance that ensure environmental protection, community rights and equitable benefit-sharing.

5.4 Recent Mining Law: Case Study Focus on Africa

The mining sector has undergone several reforms, leading to the development of new mining laws in the 2000s. For example, Burkina Faso introduced its new mining law in 2015,[38] Kenya adopted a new mining Act in 2016 and Tanzania implemented new mining laws in

[36] The 2010 Kenyan Constitution.

[37] International Institute for Sustainable Development (IISD), *How to Advance Sustainable Mining* (October 2021) https://www.iisd.org/system/files/2021-10/still-one-earth-sustainable-mining.pdf accessed 23 May 2025.

[38] In Burkina Faso, the Mining Code was adopted by the National Transition Council on 26 June 2015.

2017.[39] These laws primarily focus on attracting foreign investment while ensuring local participation and economic development from mining activities. Considering global efforts to address climate change, these mining laws also place a strong emphasis on environmental protection.

In Table 5.1, we highlight recent mining legislations and their key provisions in a few selected countries:

1. South Africa: Minerals and Petroleum Resources Development Act (MPRDA) of 2002
2. Ghana: Minerals and Mining Act, 2006 (Act 703)
3. Zambia: Zambia Mines and Minerals Development Act No. 11 of 2015
4. Uganda: Mining Act, 2003 (Cap 148)
5. Kenya: Mining Act, 2016 (No. 12 of 2016)

As illustrated in Table 5.1, most mining legislations in Africa are current and aim to address various historical challenges in the mining sector, including local content, revenue realisation and environmental protection.

Whereas national, regional and international frameworks align on key goals such as sustainability, equity and inclusivity, implementation challenges persist due to varying capacities and institutional disparities. The effective implementation of these frameworks will ensure that African countries benefit from their mineral resources. Effective implementation, however, will require targeted capacity-building, strong institutions, enhanced regional cooperation and more substantial support to ensure consistent governance standards across all levels.

5.5 Conclusion

As global demand for critical minerals grows, driven by the clean energy transition and digital technologies, ensuring their sustainable extraction has become more urgent than ever. Sustainable mining must extend

[39] Several legislations were enacted in Tanzania including the Natural Wealth and Resources (Review and Re-negotiation of Unconscionable Terms) Act, 2017; the Natural Wealth and Resources (Permanent Sovereignty) Act, 2017 and many others.

Table 5.1 Relevant mining laws in different countries

Mining law provisions	*Ghana*	*Zambia*	*Uganda*	*South Africa*	*Kenya*
Ownership of minerals	Section 1 of the mining act	Section 3—All mineral resources in Zambia are the property of the state	Section 3—All mineral resources are the property of the government	Section 3(1) —The state is the custodian of all mineral resources in South Africa	Section 4—The government of Kenya is the custodian of all minerals
Licences	Section 6 give procedure for acquiring licences	Sections 6 and 7 provides for the procedure for obtaining licences	Section 6 provides for the procedure for obtaining mining licences	Section 22—Procedure to acquiring exploration right	Section 41—Sets out a process for obtaining mining licences
Royalties	**Section 25**—Mining companies must pay royalties to the state	**Section 121**—Mining companies are required to pay royalties to the state	**Section 69**—Provides for royalties	Section 27—Royalties are payable to the state based on the value of the minerals extracted	Section 107—Royalties to be paid to the government based on the value of the minerals extracted
Local content	**Section 59**—Local labour and source goods and services from local suppliers	**Section** 77—Mining companies are required to prioritise local procurement and employment	Mining regulations provide for local content	**Section 2(1)(c)**—Promotion of local economic development	Section 56—Provides for local content
Environmental protection	**Section 71**—Environmental impact assessment (EIA)	**Sections 87–89** provides for environmental impact assessments (EIAs)	**Section 54**—The act mandates that environmental protection and rehabilitation. Plus EIAs	**Section 39**—Provides for land rehabilitate. Section 48—Provides for EIAs	Section 105—Provides for EIAs and plans for environmental management and rehabilitation

Source: Created by Authors (2025)

beyond securing resources to rigorously minimising environmental harm, safeguarding water and land ecosystems, and protecting the health and rights of local communities. This includes reducing pollution, curbing greenhouse gas emissions and enhancing safety in both large-scale and artisanal mining operations. Achieving this requires strengthening governance, enforcing robust environmental and labour standards and promoting the widespread adoption of cleaner, more efficient technologies. Ultimately, a sustainable critical minerals sector must balance resource security with social responsibility and ecological integrity, thereby supporting a just, resilient and low-carbon future.

CHAPTER 6

Towards a Global Critical Minerals Treaty

Abstract As the critical minerals industry grows in economic value and also in national strategic importance, there are global calls for a more revolutionary approach. This chapter considers one of these options and explores the reasons for a move to a new 'Global Critical Minerals Treaty'. One comparison is to examine the benefits of global cooperation on this vital sector for the global economy—this has happened more recently over the issue of 'plastics', for example. The chapter identifies how society has arrived at a key point where the critical minerals industry cannot just operate with legacy business practices from the past, and how society is at a point of revolution in the development of the sector. Strategically, the United Nations has already recognised the value of utilising principles that deliver increased just outcomes in society as appropriate for the critical minerals sector. This aligns with the perspective of this chapter and book, as well as the research perspective since circa 2020. Consequently, it is advanced here that there should be a new *Global Critical Minerals Treaty* based on principles of justice. This will deliver collectively for the mining sector's key stakeholders: governments, companies and civil society. Further, it will align the sector around other key international law and economic policies on climate change, the energy transition and the just transition.

© The Author(s), under exclusive license to Springer Nature Switzerland AG 2026
R. J. Heffron et al., *Revolution in the Global Critical Minerals Sector*, Just Transitions,
https://doi.org/10.1007/978-3-032-26366-7_6

Keywords Critical minerals • Global Critical Minerals Treaty • Mining governance • Energy transition • Climate change • Just transition • Energy justice • Regulatory frameworks • Transparency and accountability • Mineral value chains • Sustainable mining • Resource governance • Extractive industries • Policy reform • Methane and environmental impacts

6.1 The Mining Sector at a Point of Revolution

The mining sector is a growth sector of the economy for many countries, and this has been the case for over 200 years since the sector's commercialisation. There have always been issues in its growth and periods of evolution and revolution. Today, we can reflect on how there is, in effect, a revolution in the sector due to the focus on critical minerals. This latter focus stems from their contribution to the energy transition and the challenge of climate change.

Indeed, more and more of civil society are now aware of the mining sector due to these critical minerals. There is now a clear link between these critical minerals and people's everyday lives, with their use in car batteries, computers and mobile phones. There is a realisation amongst civil society that they need to have an interest in this sector as it becomes part of their daily livelihood. There are key questions around how the key issues of responsibility, risk and reward (known as the 3R's),[1] are shared within the sector among the three key stakeholders: government, the private sector (companies) and civil society.

Civil society now wants to take a further step and engage with critical minerals and, therefore, the mining sector. This was not to the same level as before until critical minerals became more prominent in the last five years. Hence, there is a more revolutionary phase of the mining sector, requiring a new set of guidelines for its operation. The old frameworks are no longer what 'best practice' should be, and unquestionably many would say they were not working.

[1] Heffron (2024) refers to these as the 3R's (Risk, Reward and Responsibility)—of project development—and particularly important for the mining sector: Heffron, R. J. 2024. Energy Justice—The First Step in an Energy Decision Today. Heffron, R. J. and Fontenelle, L (editors). 2024. The Power of Energy Justice & the Social Contract. Springer: Heidelberg, Germany.

This chapter highlights the major theme within the book that, with the advent of critical minerals, the mining sector is at a strategic point in its development. This section explains the reasoning behind this before Sect. 6.2 examines why the next steps should be supported by a framework, toolkit and principles of justice. Then, Sect. 6.3 concludes the chapter and the book by exploring in brief why the call for a new *Global Critical Minerals Treaty* has arisen.

Accidents, major pollution, environmental impacts and carbon dioxide emissions have long plagued the mining sector. While these have been documented in many countries, there remain a wide range of undocumented issues that are not recorded for a variety of reasons.[2] The criticism of how successful the governance, laws, economics and policy of the mining sector could be is far higher than it is. However, civil society, unless directly affected (by living close to a mine, for example), has remained largely detached from how the mining sector operates.

This is no longer the case and there is increased interest in how the mining sector operates with the advance of the importance of critical minerals. More knowledge of its problems, its operations and its impacts is now visible and civil society is taking a more active role.[3] Pressure lies on governments and companies to be more transparent. There have even been successful international initiatives in this context with the launch of the Extractives Industry Transparency Initiative (EITI).[4] In many coun-

[2] There are multiple accounts of this but see the following as being three comprehensive studies: (1) Macklin, M. G. et al. 2023. Impacts of metal mining on river systems: a global assessment. Science, Vol 381 (6664) 1345–135; (2) United Nations Environment Programme. 2024. Climate Risks in the Metals and Mining Sectors. Available at: https://www.unepfi.org/wordpress/wp-content/uploads/2024/05/Climate-Risks-in-the-Metals-and-Mining-Sector-1.pdf (last accessed 8 June 2025); and (3) Oxfam International. 2021–22. Oxford Natural Resource Justice Program Report. Available at: https://express.adobe.com/page/WL3z3jwjIXITE/ (last accessed 8 June 2025).

[3] Again here there are multiple examples here and we just select several of these: (1) Zuka, S. P. 2023. Non-governmental organisations and the politics of mining law review in Malawi: Bifurcations, subjectivities and contested emancipation agenda. The Extractive Industries and Society, 14, 101,252; (2) Oxfam International. 2021. In Ghana Civil Society is driving Mining and Oil Reforms. Available at: https://politicsofpoverty.oxfamamerica.org/in-ghana-civil-society-is-driving-mining-and-oil-reforms/ (last accessed 8 June 2025); and (3) Aspinwall, M. 2021. Bringing rights to life: How civil society organizations help guarantee participation rights in developing countries. The Extractive Industries and Society, 8 (3), 100,923.

[4] See the website for information on this international initiative - Extractive Industries Transparency Initiative - available at: https://eiti.org/ (last accessed 8 June 2025).

tries, despite the introduction of legislation on environmental impacts (environmental impact assessment legislation), progress overall has been mixed.[5] Further, many countries continue to wait for international tax reforms[6] and how they have aim to improve the distribution of revenue in the mining sector. Regardless of the evidence of progress or constant evolution, revolutionary change is what is needed.

Today, there are increasing calls for change[7] which take the form of increased legal action, withdrawal of public support and social unrest. Thus, social unrest is fast becoming one of the main reasons for the delay or abandonment of mining projects.[8] The relationship between key stakeholders in many countries is in a state of flummox, with their expectations unaligned. However, expectations of civil society and government are on the rise, particularly regarding financial issues. With the growth and future potential growth of the industry, there is an expectation that more of the financial benefits from the industry should be shared amongst the stakeholders, but that struggle remains.

A key debate has emerged around improving 'equity' in the mining sector in the last five years.[9] More broadly, this is referred to as justice, which includes ensuring a general level of fairness and inclusivity (ensuring that no one is left behind—which has also given rise to the just transition). The underlying objective is to ensure the fair distribution of the benefits and the negatives of the mining sector. However, the prevailing view is that civil society bears a disproportionate share of the negatives while rarely

[5] UN Environment. 2018. Assessing Environmental Impacts A Global Review of Legislation. Nairobi, Kenya.

[6] IISD (Readhead, A). 2018. Tax Incentives in Mining: Minimising Risks to Revenue. https://www.oecd.org/content/dam/oecd/en/about/programmes/beps-in-mining/tax-incentives-in-mining-minimising-risks-to-revenue.pdf (last accessed 8 June 2025).

[7] World Economic Forum (Aguilar, A). 2025. Mining's next chapter: driving innovation, resource stewardship and global progress. Available at: https://www.weforum.org/stories/2025/01/mining-innovation-resource-stewardship-global-progress/ (last accessed 8 June 2025).

[8] Global Witness. 2024. Critical mineral mines tied to 111 violent incidents and protests on average a year. Available at: https://globalwitness.org/en/campaigns/transition-minerals/critical-mineral-mines-tied-to-111-violent-incidents-and-protests-on-average-a-year/ (last accessed 8 June 2025). And see some academic research in the area too on social protests and their high risk: Teschner, B. and Holley, E. The cost of mine suspension from social conflict: A decision tree model. Resources Policy, 74, 101,443.

[9] *Heffron, R. J. 2020. The Role of Justice in Developing Critical Minerals. The Extractives Industry and Society, 7 (3), 855–863.*

receives any benefit, given the many challenges facing the mining sector today.

Despite a range of different and short perspectives on improving justice in the mining sector by different organisations, many have lacked a more systematic and holistic approach. In academic literature, research has been pushed in this direction with many examples.[10] In the policy area, however, the United Nations has identified a clear path forward with an explicit focus on justice and equity in the critical minerals sector.[11] This is a significant development because the mining sector has struggled for years to secure international policy action (despite the success of EITI).

The UN report entitled *Resourcing the Energy Transition: Principles to Guide Critical Energy Transition Minerals Towards Equity and Justice* is based on seven principles. These principles are more action-based with actionable points, but they have their roots in the justice principles outlined earlier in Chap. 2, see in Table 6.1. Together, these UN-supported principles with these justice principles can provide the framework for a *Global Critical Minerals Treaty*.

This book aligns with the work of the UN and aims to provide further insight into the issue of justice in the mining sector and for critical minerals. It follows a more research-led perspective, and highlights the key justice issues from an interdisciplinary perspective. There is a principle-based approach to ensuring the practice of justice within the mining sector. Countries worldwide, including Colombia, the United States and South

[10] There are examples from regions (Qurbani, I. D., Heffron, R. J. & Rifano, A. T. S. 2021. Justice and critical mineral development in Indonesia and across ASEAN. The Extractives Industry and Society, 8 (1), 355–362), specific topics (*corporations* - Heffron, R. J. 2023. Repurposing for the just transition: energy companies need to future-proof their structure and strategy. Journal of World Energy Law and Business, 16 (3), 302–307; *sovereign wealth funds* - Heffron, R. J. 2018. The application of distributive justice to energy taxation utilising sovereign wealth funds. Energy Policy, 122, 649–654; and the *social license to operate* - Heffron, R. J., Downes, L., Rodriguez, O. & McCauley, D. 2021. The emergence of the 'social licence to operate' in the extractive industries? Resources Policy, 74, 101,272.) and then the following details some of the key literature on mining and justice (Heffron, R. J. 2020. The Role of Justice in Developing Critical Minerals. The Extractives Industry and Society, 7 (3), 855–863).

[11] United Nations. 2024. Principles to Guide Critical Energy Transition Minerals towards Equity and Justice Available at: https://www.un.org/sites/un2.un.org/files/report_sg_panel_on_critical_energy_transition_minerals_11_sept_2024.pdf (last accessed, 8 June 2025).

Table 6.1 UN and Justice Principles Relationship

No.	*UN principle*	*The five justice principles: distributive, procedural, restorative, recognition and cosmopolitan*
1	Human rights must be at the core of all mineral value chains	Involves all five justice principles
2	The integrity of the planet, its environment and biodiversity must be safeguarded	Restorative and cosmopolitan justice
3	Justice and equity must underpin mineral value chains	All 5 justice principles
4	Development must be fostered through benefit sharing, value addition and economic diversification	Distributive and recognition justice
5	Investments, finance and trade must be responsible and fair	Distributive justice and cosmopolitan justice
6	Transparency, accountability and anti-corruption measures are necessary to ensure good governance	Procedural justice
7	Multilateral and international cooperation must underpin global action and promote peace and security	Cosmopolitan justice

Source: Created by Authors (2025)

Africa, have started to operationalise this approach in their respective energy sectors.[12]

At this point in 2025, society has reached a critical point of injustice which demands more just outcomes from mining operations. There is no more room for inaction, naivety or unawareness of essential justice issues. The knowledge is readily available, and companies should be held more accountable in their decision-making now and in the future. Indeed, decision-makers across the mining sector have access to more knowledge than ever before and must therefore be held to higher standards of transparency and accountability. Particularly, it should be expected that they have greater knowledge on taxation, environmental impact and climate challenges.

[12] Heffron, R. J. 2024. Energy justice – the triumvirate of tenets revisited and revised. *Journal of Energy and Natural Resources Law*, 42 (2), 227–233.

6.2 The Role of Justice in the Future of the Mining and Critical Minerals Sector

The UN report clearly demonstrates that justice has a central role in the mining sector.[13] At its core, the report highlights the importance of accountability and transparency in decision-making. Accordingly, decision makers—whether policymakers, company executives or community leaders—must demonstrate how they aim to incorporate 'justice and equity' in their decision and actions. Accountability and transparency in decision-making now include this element of justice as a requisite issue for evaluation.

The UN also advocates some actionable areas based on their principles. While these are valuable, they could adopt a more comprehensive approach. For example, Table 6.2 highlights the UN actionable points and their link with the five justice principles.

It is necessary, therefore, to consider in more detail what type of justice and equity framework a decision-maker should consider including in their analysis. Based on over 10 years of energy justice-related research, we adapt and advance a Justice Milestone Reporting framework proposed by Heffron et al. (2024), published in the prestigious *Nature Energy* journal. The heart of what is advanced there is a justice toolkit which can be applied to the mining and, more specifically, the critical minerals sector. The adapted toolkit for the development of critical minerals is presented in Table 6.3.

The advantage of this approach—the toolkit for Energy Justice Milestones Reports—is that it ensures the retention of justice and equity throughout the decision-making process. The aim is that the incorporation of justice and equity into decision-making is not just in a single situation but it is a periodic approach with multiple points where justice and equity will be considered. The example that can be considered is a mining project comprising four stages: planning to construction to operation to decommissioning. Therefore, based on the principle of restorative justice, there are justice considerations involved at the planning phase in

[13] United Nations. 2024. Principles to Guide Critical Energy Transition Minerals towards Equity and Justice Available at: https://www.un.org/sites/un2.un.org/files/report_sg_panel_on_critical_energy_transition_minerals_11_sept_2024.pdf (last accessed, 8 June 2025).

Table 6.2 UN actionable recommendations

No.	*UN actionable recommendations*	*The five justice principles: distributive, procedural, restorative, recognition and cosmopolitan*
1	A high-level expert advisory group to accelerate greater benefit-sharing, value addition and economic diversification in critical energy transition minerals value chains, as well as responsible and fair trade, investment, finance and taxation	Procedural justice
2	A global traceability, transparency and accountability framework along the entire mineral value chain—From mining to recycling—To strengthen due diligence, facilitate corporate accountability and build a global market for critical energy transition minerals, though the framework should not be used as a unilateral trade barrier	Procedural justice with elements of distributive and cosmopolitan justice
3	A global mining legacy fund to build trust and address legacy issues as a result of derelict, ownerless or abandoned mines and strengthen financial assurance mechanisms for mine closure and rehabilitation	Restorative justice with elements of distributive justice
4	An initiative that empowers artisanal and small-scale miners to become agents of transformation to foster development, environmental stewardship and human rights	Recognition, distributive, procedural and restorative justice
5	Equitable targets and timelines for the implementation of material efficiency and circularity approaches across the entire life cycle of critical energy transition minerals	Distributive, recognition, procedural and cosmopolitan justice

Source: Created by Authors (2025)

environmental impact assessments, and the development of a clean-up fund in the operation phase, for example.[14]

By following such a toolkit, more just outcomes will occur in the mining and critical minerals sector. Stakeholders can collaborate in the collective delivery of justice by strengthening their relationship with each other over time. All stakeholders must accept that justice will not be immediately achieved in every project due to different variables and journeys, and in essence, different justice pathways will eventually emerge. As projects

[14] Heffron, R. J. and Hazrati, M. 2024. The Rise of Restorative Justice in the Energy Transition and for Climate Mitigation. Springer: Heidelberg, Germany.

Table 6.3 The toolkit for energy justice milestones reports

Energy justice principle	*Key areas to cover in each report*
Procedural	Review of the legal architecture Transparency and participation procedures Periodic review of procedural justice Performance during project lifecycle
Distributive	Financial impact of the project Taxation roadmap Job creation and innovation Challenges to fair distribution
Restorative	Environmental impact assessment Compliance Development of clean-up fund resolving legacy energy waste prevention roadmap
Recognition	Stakeholder identification (mapping) Local capacity development Tracking respect for human rights Legacy for future generations
Cosmopolitan	Comparison with alternative projects Cross-region/border impacts Obligation to other countries International obligations (including financial)

develop, what is considered 'best-practice' will evolve, and model policies on justice and equity for mining operations will continue to be refined.

As recognised by the UN, immediate action is required in this area. Many other international organisations also recognise the need for action, and there is an increased number of reports available on critical minerals.[15] The growth of the industry, as stated in earlier chapters, is clearly evident. In some countries, growth could even be faster with straightforward processes to follow for the development of these critical minerals in particular—i.e., investment frameworks. Also, as noted in earlier chapters, while there are new laws and policies to facilitate the development of critical minerals, there are emergent problems in this area.

[15] Nearly every international NGO, and international organisation has produced a report on critical minerals within the last 3 years if not before.

Some new laws are limited in their focus, notably the EU Critical Raw Materials Act.[16] The legislation makes no mention of the concepts of 'justice' or 'equity'. Instead, it focuses primarily on development and aligns with other EU law that may contribute to more just outcomes. Overall, however, the EU approach does not reflect the UN's more recent policy perspective, and it is advanced here that the EU legislation is already outdated in this regard. This evinces a familiar tale in the mining sector, where legislation often fails to keep pace with industry developments. The EU has adopted a slow evolutionary approach to regulating the mining sector rather than a revolution. This mirrors the EU policy action on battery production and the overall progress on green technology production. This is illustrated by the European (particularly German) automobile industry, which has largely responded to market developments through reactive measures in terms of innovation, and calls for 'protectionist' policies to be adopted for the industry as opposed to sustained innovation.[17]

Other regions of the world have also been active, with critical minerals strategies advanced in a number of areas, from Africa to Asia. The success of these has been to illuminate that law and policy action should happen. However, countries are still making slow progress in developing it and creating conditions that encourage investment, build internal tax revenue and benefit civil society. The longstanding challenges facing the mining sector— persisting for decades, if not nearly two centuries— have either re-emerged or, depending on one's perspective, never truly disappeared.

The question for the future remains: *why is legislative and policy change so slow in the mining sector?* Further, *why do we ask the same questions on mining in a 200-year-old industry?* Child miners were working down the coal mines in the 1800s, and till date, society has the same challenge. Though some would cite the issue of the resource curse,

[16] European Critical Raw Materials Act. 2024. Available at: https://single-market-economy.ec.europa.eu/sectors/raw-materials/areas-specific-interest/critical-raw-materials/critical-raw-materials-act_en (last accessed, 8 June 2025). In full it is also available here: https://eur-lex.europa.eu/legal-content/EN/TXT/PDF/?uri=OJ:L_202401252 (last accessed, 8 June 2025).

[17] European Parliamentary Research Service. 2024. The crisis facing the EU's automotive industry. Available at: https://www.europarl.europa.eu/RegData/etudes/ATAG/2024/762419/EPRS_ATA(2024)762419_EN.pdf (last accessed, 8 June 2025). And see also an industry view: McKinsey. 2025. European automotive industry: What it takes to regain competitiveness. Available at: https://www.mckinsey.com/industries/automotive-and-assembly/our-insights/european-automotive-industry-what-it-takes-to-regain-competitiveness (last accessed, 8 June 2025).

the overarching problem is one of slow progress in the legislative and policy change for mining. Another reason might be the influence of the present multinational corporate interests (previously more national or state-owned companies) with little to no action to curb this (despite progress around the EITI). Overall, the central question as to *why legislative and policy change is so slow in the mining sector* remains an open question.

6.3 A Global Mining Vision for Critical Minerals: A New Global Critical Minerals Treaty

The reasons for inaction in developing new mining law and policy are multi-layered. The key reason lies in failure to recognise is the need for a revolution in mining law and policy, resulting in missed opportunity. For many countries, this is an economic opportunity which could bring significant direct and indirect economic benefits to growth-starved countries. To continue on the existing pathway is to play and operate as before, an approach which has neither attracted sufficient investment nor delivered equitable benefits to many governments and civil societies.

There is the obvious concern in many countries that too much law and policy will discourage investors due to rigidity or cumbersome processes. This view, in effect, causes a '*race to the bottom*' in terms of taxation,[18] environmental performance and implementation of standards or principles in operations. However, this is not often the case; the economic case for extraction must be properly assessed, allowing greater inputs from local through to national governments. In many instances, the resources are extracted because they are needed, and future values will reflect this continued demand. For example, extraction happens at lower financial margins in developed countries, where regulatory standards are generally more stringent.[19]

Indeed, one of the concerns for the United States (prior to the new Trump administration) had been whether US companies had to comply

[18] Multiple accounts of this area available but here are two relevant for our chapter: (1) Tax Justice Network. 2020. Tax competition and the race to the bottom. Available at: https://taxjustice.net/topics/tax-competition-and-the-race-to-the-bottom/ (last accessed, 8 June 2025); (2) Bharanidaran. R. 2024. Ending the Race to the Bottom: Analyzing A Recent Global Agreement on Corporate Taxation. Chicago Journal of International Law. 3 (1) 37–69.

[19] Well that is the stated view but many researchers would say that the practice and implementation of mining law in developed countries is also lacking.

with higher standards of operation than some of its competitors in the growing critical minerals industry. The question arose as to whether such higher standards could provide US companies with a medium- to long-term competitive advantage. The answer will probably remain uncertain until post-Trump administration, and, in any case, it is unlikely that US companies will be able to lower their operational standards. In addition, with agencies like USAID and other international action now unfunded, many US companies will have lost access to significant information on the actions of competitors.

Overall, however, the mining sector has always had a simple, naïve and uncomplex objective in society. Its presence was always economic with an aim of profits for industry, potentially some tax revenue for government and jobs for local/national populations. The amount of profits or revenue and the quality of the jobs created were all developed in a framework of expectation (which we can refer to as the 'mining mirage'). Non-performance or manipulation of this framework of expectation has resulted in the mirage of mining booms[20] continuing across the world.

The legacy effect of past mirages of mining booms is evident all across the world in multiple forms: (1) from ones that never happened; (2) to ongoing environmental impacts from mines in operation; (3) job growth that remain non-existent as foreign workers fill the job requirements; (4) jobs created are indecent, low paid, and short-term; (5) environmental destruction at the mining site with old sites left unrestored; (6) no clear evidence of tax revenue and its domestic retention; (7) lack of coherent community benefits programme; and (8) unclear long-term benefits from a project to a country.

In order for the critical minerals mining boom to have a different pathway, a revolutionary approach to law and policymaking for the industry is needed. The UN publication[21] is a step in this direction. Different regions have equally approached it from a regional perspective, for example, the

[20] Overseas Development Institute. 2024. The 'strategic mirage' of Africa's green minerals wealth. Available at: https://odi.org/en/insights/the-strategic-mirage-of-africas-green-minerals-wealth/ (last accessed, 8 June 2025); and International Institute for Sustainable Development. 2016. Mining a Mirage? Reassessing the shared-value paradigm in light of the technological advances in the mining sector. Available at: https://www.iisd.org/system/files/publications/mining-a-mirage.pdf (last accessed, 8 June 2025).

[21] United Nations. 2024. Principles to Guide Critical Energy Transition Minerals towards Equity and Justice Available at: https://www.un.org/sites/un2.un.org/files/report_sg_panel_on_critical_energy_transition_minerals_11_sept_2024.pdf (last accessed, 8 June 2025).

Africa Mining Vision. Meanwhile, it is questionable whether these have sufficient ambition. Though the next aims for the UN vision remains unclear, there are calls for a new global approach and even a global critical minerals treaty. There are clear benefits to such an international treaty considering the importance of these critical minerals in contributing to economic stability for many countries in the future. In addition, for many other countries, they offer a new way of delivering additional improvement to their economy, particularly in developing countries.

Early research or views on a *Global Treaty for Critical Minerals* propose that transparency and traceability are vital.[22] These calls are all valid and, in particular, some raise important issues related to justice. Significantly, these debates took place at UN COP30 in 2025, where critical minerals were highlighted alongside the annual energy, climate and just transition debates. Unfortunately, critical minerals were removed from the final text during the closing stages of the negotiations because of concerns over their link to governance, with several countries opposing their inclusion. The essential requirement are clear principles that guide the creation of such a Treaty. A new global vision for mining has to envisage a critical minerals industry that delivers benefits (whether they be short term, medium term or long term) for all the key stakeholders, such as the governments, companies and civil society. The aims must ensure that no one is left behind in the future development of the mining sector. This approach over time has less risk for all stakeholders and can be accelerated by more collective action—such as through a global treaty.

A global mining vision and treaty can build on the UN policy initiative. It is a growth area for future research and policy practice. The work of the UN could be more ambitious, with some of the principles aiming to ensure they align with other key UN policies on climate change, energy and the just transition. Identifying how additional mechanisms of accountability and transparency can be built into business operations will also be valuable. In this context, how data is gathered, maintained and utilised is needed for mining operations. The lack of such 'mining data' places

[22] Business and Human Rights Resource Centre. 2024. Colombia proposes a binding treaty for responsible mining of critical minerals at COP30Available at: https://www.business-humanrights.org/fr/derni%C3%A8res-actualit%C3%A9s/colombia-and-brazil-propose-a-binding-treaty-for-responsible-mining-of-critical-minerals-at-cop30/ (last accessed, 8 June 2025).

different stakeholders at a disadvantage, ultimately affecting the development of a just, equitable and sustainable critical mineral mining sector.

There are different possible pathways forward for the mining sector with the advent of critical minerals. However, in a few years, we should not be looking at an industry that only aspires to create problems associated with past legacies of operations and practice. Instead, a new revolutionary approach to mining law and policy is needed. The initial first steps to a *Global Treaty for Critical Minerals* should be the aim. As this industry develops, it should align with principles of justice that can deliver improved just outcomes for all society's stakeholders—governments, companies and civil society.

Index[1]

[1] Note: Page numbers followed by 'n' refer to notes.

© The Author(s), under exclusive license to Springer Nature Switzerland AG 2026

R. J. Heffron et al., *Revolution in the Global Critical Minerals Sector*, Just Transitions,
https://doi.org/10.1007/978-3-032-26366-7